The Vanishing Aromanian

EUGENE MATZOTA

THE VANISHING AROMANIAN, by Eugene MATZOTA

Cover Design: Eugene Matzota
Interior Design: Eugene Matzota
Book Editor: Monica Szemethy
Publisher: CreateSpace, a DBA of On-Demand Publishing, LLC

ISBN-13: 978-1984933997

ISBN 9781984933997

90000 >

9 781984 933997

ISBN-10: 198493399X

THE VANISHING AROMANIAN, 2018
First Edition

The Vanishing Aromanian

EUGENE MATZOTA

2018

CONTENTS

ABOUT THE AUTHOR 11

FOREWORD 13

A-Romanians, Romanians? 16
Getting back to the roots 18
The Aromanian Question, a taboo subject? 19

STATUS QUO 23

So many names, a unique fate 27
Community, ethnicity, nation.................................. 28
One language, one nation, one state? 29
Pelasgians and Aromanians 31

JUST A KIND OF GLOSSARY 37

VLACH or AROMANIAN 39
VLACH or MACEDONIAN 41
VLACH or VALACH (WALACH) 43
ABELA, not AVDELLA.. 49

A HISTORY OF HISTORIES 55

Why necessarily a history of histories? 57
Mythistory, myth and/or history? 59

The historian, patriot or traitor? *60*

The history of myth or the myth of history? *64*

'POLITICALLY CORRECTNESS' AND HISTORY 67

That lie called 'politically correct' *69*

A world only for the Aromanians *71*

'Politically correct' Latinization *73*

Then, where are the Aromanians coming from? *77*

WHO ARE THE AROMANIANS? 87

Where is the country of the Aromanians? *89*

What have Aromanians done for the others? *92*

Aromanians, people like us *93*

Aromanians and their social values *95*

Where do the Aromanians' name come from? *97*

Aromanians, religion and Church *99*

DNA, THE FINAL ANSWER? 103

Does DNA really have all the answers? *105*

DNA tests conclusions *113*

FACTS AND FIGURES - BEFORE 1800 117

First written evidence of the Aromanian language
.... *119*

The first mention of the ethnonym 'Vlach' *123*

Vlachs and Macedonians, clearly different *124*

Kekaumenos about the Vlachs in Thessaly *125*

'Alexiada' and the Vlachs *125*

Vlach families on Mount Athos *127*

The rebellions of the Vlachs *127*

Great Wallachia *128*

Despotate of Epirus *129*

The Chrysostom of Stefan Prvovencani *130*

The Vlach-Bulgarian Tsardom *131*

The resettlement Vlachs of Thrace *135*

Descriptio Europae Orientalis *136*

Great Vlachia *138*

Stefan Dusan's Code *139*

Vlach special rights *139*

Ragusa calls the Vlachs *140*

Laonic Chalkokondyl on Vlachs *140*

Vlachs in Hungary, at Miskolc *141*

Statuta Valachorum *141*

Vlachs in 'De regno Dalmatiae et Croatiae' *142*

Vlachs and Europe *145*

Maria Gianni Manarutta and Vlachs in Trieste... *145*

Cantemir and Vlachs *146*

Trading company in Miskolc *147*

The Academy of Moscopole *147*

The destruction of Moscopole *147*

Emigration to the West *149*

Miron Costin and the Vlachs *150*

FACTS AND FIGURES - AFTER 1800 151

The Dictionary of Daniil Moscopoleanul *153*

Aromanian Grammar of Mihail Boiagi *154*

The Aromanians of Tudor Vladimirescu *158*

Memories and manifests *161*

First Macedo-Romanian Committee *162*

The first Romanian school in Turkey *162*

The Great Vizier against the Greek Patriarchate *165*

The Macedo-Romanian Culture Society *166*

The Romanian High School in Bitolia *167*

The Greek Patriarchate fights the Aromanians ... 169
The situation is getting worse................................ 170
The Irade dedicated to Aromanians...................... 171
The bizarre position of Romania........................... 172

BALKAN WARS 173

The outbreak of conflict...................................... 175
Aromanian Question 182
Romania's demands for Aromanians 183
The Peace of Bucharest, 10 August 1913.............. 190
Everybody wins? Well, not exactly... 193

AFTER THE TREATY..................................... 199

Principality of Pindos... 201
Peace with Bulgaria.. 203
Greco-Turkish war.. 204
Little Entente (1920 – 1921)................................. 204
The Communists and 'the Aromanian Question'. 207

GREAT AROMANIANS - BEFORE 1800........ 209

Nicoliță, head of the Vlachs 215
St. Nicodim of Tismana 216
Chalkeus, the great figure of the Enlightenment.. 217
First book in the Aromanian language 217
Constantin Ucuta and 'New Pedagogy'................ 218
Metropolitan Filaret... 219
St. Joseph the New of Partos................................ 219
Velestinlis, the father of the modern Ellada......... 220

GREAT AROMANIANS - AFTER 1800 221

Coletti, the ideologist of the Great Byzantium 223
The Grammar of Mihail G. Boiagi........................ 224

Gheorghe Roja *225*

Emanuil Gojdu, the great fighter *225*

Andrei Șaguna, the Metropolitan *230*

Panu and Romania's strategy for the Balkans *234*

Lefkada, a great poet of modern Greece *234*

Ion Ghica, academician and Prime Minister *235*

Mărgărit, apostle of Romanian spirituality *235*

Ioan D. Caragiani and the Romanian Academy... *237*

Belimace, author of the Aromanian Hymn *238*

Spyridon Lambros, Prime Minister *239*

Kostas Krystallis, Greek writer and poet *239*

Asdreni, important Albanian poet *239*

The Manachea brothers (Manaki) *240*

GREAT FAMILIES **245**

Mocioni family *247*

The Sina Family *252*

The Dumba Family *256*

THE VANISHING AROMANIAN? **263**

Romania and the vanishing Aromanian *265*

Aromanians and Romanians today *268*

The new Millenium Aromanian *269*

ANNEXES **273**

Shared admixture events in Eastern Europe in the first Millenium *275*

Die Aromunen. Ethnographisch-philologisch-historische Untersuchungen: Vorrede. *279*

Voyage de la Grèce, Pouqueville *282*

Descriptio Europae Orientalis *286*

The Romanians of the Balkan Peninsula *291*

Instructions given by Titu Maiorescu.................... 297
Pro-memory submitted to the Foreign-Office 301
ROUMANIE, GRECE, MONTENEGRO, SERBIE,
BULGARIE.... 304
Traité de paix signé à Bucarest........................... 304
The description of Nikolaus Dumba..................... 312

BIBLIOGRAPHY 315

BOOKS... 315
WEB.. 321

ILLUSTRATIONS... 324

ABOUT THE AUTHOR

Member of the Theosophical Society in Adyar, India, since 1997. Author of many books in the field of Theosophy.

Eugene Matzota is a deep and multifaceted personality, engaged in many areas, believing that his mission in life is to provide knowledge to those seeking a way into the transcendent.

His care for the highest mystical aspiration is just an ambition for higher spiritual knowledge gained by spiritual experience. This quest for spirituality, originally started only with the naive desire of the impetuous adolescent who wants to know everything, has become his attitude to the world.

He published in 2012 the only book that presents only facts and facts about Aromanians, a quick guide for those who will give, not comments on Aromanians: History of Aromanians in Data.

Born and raised in Oradea, he has studied at the Emanuil Gojdu National College, a great Aromanian. He found out later on that he came from an old family, related to the founders of the Epirus Despotate (Principality) some eight centuries ago.

And so he realized why he held other values than the others. He understood why he still believes in ideas such as those of honor, character, and chivalry. Why is he so different...

Regaining his own Aromanian tradition made him confident in his values, in his direction in life.

FOREWORD

'To which family should one connect the inhabitants of ancient Dacia with those of the Vlach tribes of European Turkey?

Do they belong to the ancient nationality of the Pelasgians?

Do they descend from Roman settlers?

An agglomeration of people pulled in from the four corners of Europe, would it have a nationality?'[1]

---

[1] A quelle famille faut-il rattacher les habitants de l'ancienne Dacie et ceux des tribus valaques de la Turquie d'Europe? Appartiennent-ils à l'antique nationalité des Pélasges ? Descendent-ils des colons Romains ? Une agglomération d'hommes tirés des quatre coins de l'Europe, aurait-elle une nationalité?
Colson F., *Nationalité et régénération des paysans moldo-valaques*, E. Dentu, Paris, 1862, p.13

A-Romanians, Romanians?

I have to say from the very beginning that this is a book that may be written unwittingly from a Romanian point of view, because the author was born in Romania. On the other hand, the author is coming from a family with 800 years of Aromanian history.

Therefore, I must have a better understanding of the problems related both to Romanians and to Aromanians. After years of study, I have serious reasons to believe that Aromanians are not Romanians, as the state policy of Romania tries to impose for the more than a century, but rather brothers. And this is not at all a blasphemy...

Now, for somebody who has no idea where the Balkans could be on the world map, this word, *Aromanian*, maybe doesn't mean a thing. *Aromanians, A-Romanians*, and more than this, *Vlachs*? Too much confusion here...

What could this term, *Aromanian*, mean? Probably, and this could be the first guess, these *Aromanians* are not *Romanians*, as the name might suggest, compared to other words like *symmetry* and *asymmetry*[2].

First of all, let's see what *Encyclopædia Britannica*, namely *Prof. Victor A. Friedman,*[3] has to say about this:

---

[2] A is prefix meaning "not," from Greek a-, an- "not."
[3] Andrew W. Mellon Distinguished Service Professor, Dept. of Linguistics, University of Chicago, and Director, Center for East European and Russian/Eurasian Studies

> *Vlach, also spelled Vlah, autonyms Armãn and Rãmãn, also called Aromanian, Macedoromanian, and Macedo-Vlach, any of a group of Romance-language speakers who live South of the Danube in what are now Southern Albania, Northern Greece, the Republic of Macedonia, and Southwestern Bulgaria. Vlach is the English-language term used to describe such an individual. The majority of Vlachs speak Aromanian, [...]*[4]

Well, if you try to find something about this in *Larousse*, the answer is much shorter and without any doubt, *the Aromanians are Romanians*:

> *Aroumains - Population d'origine roumaine, qui a émigré aux IXe-Xe s. dans la péninsule balkanique et vit aujourd'hui en Grèce, en Serbie-et-Monténégro et en Albanie.*[5]

One can't find nothing about this word, *Aromanian,* in *Merriam-Webster*, only *Vlach*, therefore this is what you may find there:

---

[4] https://www.britannica.com/topic/Vlach
[5] http://www.larousse.fr/encyclopedie/divers/Aroumains/
106205#iaQiaxZqf2Lkoicy.99

A member of a people scattered through Southeastern Europe originating in the early middle ages probably in the Balkans, speaking a Romanian dialect, and including chiefly mountain herdsmen (as in Northwestern Greece) — called also Wallach, Wallachian.[6]

Getting back to the roots

Therefore, after trying in vain to find an answer about the reason for the presence of this letter *A* before the word *Romanian*, I have decided to write this book...

Well, this is not a new idea, since it occurred to me when I found out that my Romanian family had a really long Aromanian history, being one of the founding families of the Despotate of Epirus some 800 years ago.

Now, having this confirmation from a man whom I know and cherish a lot, namely *Branislav Stefanoski - Al Dabija*, a fundamental change in my life did occur, giving me the real meaning of all those memories of my family, particularly the way I was educated at home, with my family using a very different set of values from the rest of the people around us.

Unfortunately, this happened only some years ago and the first thing I did as an Aromanian, was to translate and publish a challenging book written by Stefanoski - Al Dabija: *'Short descriptive history about the origin of the Arm'n–Macedonians (From pre-history to the colonization of Dacia)'.*

---

[6] https://www.merriam-webster.com/dictionary/Vlach

Later on, gathering enough material to publish all by myself a challenge to all those books about Aromanians written before me, I managed to publish in 2012 'The history of Aromanians in data', the first book containing only data and facts about 'the Aromanian issue', 'Die aromunische Frage', as *Max Peyfuss* said.

Being then aware of that real need for information, data and facts, not only opinions, I've begun to write a more complex work, based in part on what was well-structured in 'The History of Aromanians in Data'. Where there was nothing important to add or to change, I've just preserved ideas as they were, but only after discussing with experts in Aromanian issues.

The Aromanian Question, a taboo subject?

This book does not avoid some taboo problems that are not 'politically correct', well hidden under the carpet. This damaging attitude does not solve any problems; it only keeps them intact, any approach being treated as taboo.

I realized during the last ten years that the problems should be treated differently. Nowadays, people have no time to waste, to read scientific works with so many quotes over quotes and notes over notes. They want to know quickly, right now, if possible!

These considerations have led to a version that includes the main problems, trying in the meantime not to hinder reading by using too many quotes and notes, which would otherwise give so much credit to a scientific paper. Because this book is written mainly for the Romanians that still don't understand the Aromanian

Question, the chosen title is clear and direct, too, with a title that clearly suggest that Aromanians are not Romanians: *'Romanians vs. Aromanians?'*

Acad. Ioan Aurel Pop, a great open-minded Romanian historian, Rector of Babes-Bolyai University of Cluj-Napoca, said in 2016, at the First National Congress of the Romanian historians, that those who want to write about the past, but are not historians by profession, can do this provided that they *'specify the nature of the essay, the story, the impression of their products.'*

This is exactly what I do here, compelled by the silence of the historians, from the position of somebody who used to be an investigative journalist, studied and wrote about Romanian and Aromanians and, not least, knows something about addressing today's politicians.

STATUS QUO

- *The imminent disappearance of the Aromanians' identity, who are brothers with the Romanians, not more, and do not have the same origin and identity;*

- *There is too much nationalism in the Balkans;*

- *The Balkans still are the powder keg of Europe;*

- *Dividing the Ottoman Empire in the early twentieth century into small worlds, with new nations, with new targets and new nationalist leaders, led and still lead to the struggle for land and against the 'others';*

- *Each Balkan people have only a few friends, but many enemies.*

So many names, a unique fate

Yiani *Mantsu*, speaking in the name of the international organization *Consillu Makedonaromânjloru*, said in 2013, in Tirana,[7] that the Aromanians are known in Romania as *aromâni*, and in other languages as: *Aromanian, Aroumain, Aromune, Arumano, Arumeno* etc.

The Aromanians are called *Vlach (Vlleh / 'Çoban')* in Albania, *Vlasi*, in Macedonia, *Vlach (Hellino-Vlach)*, in Grecia, *Cincar (Zinzar)* in Serbia and Bulgaria, and *Mazedo-Romanen* in Germany.

They call themselves *Ar'mân / Makedon-Ar'mân*, a name identifying the native space, ancient Macedonia, says Mr. Mantsu. Even in Romania, before *Gustav Weigand* coined the term *Aromanian* in 1895, the common term was *macedo-român (Macedo-Aromanian)*,

---

[7] Mantsu, Yiani, *Juridical and political aspects regarding the minority of the Makedon-Armâns in Albania and how they enjoy all the rights granted by the European and international norms*, Conference on Minorities, Tirana, 2013

used also in the title of the *Macedo-Romanian Cultural Society (Societatea de Cultură Macedo-Română)*, established in Romania in 1879. Another term used in those days was *Macedonian*.

We will use here *Aromanian*, just to simplify things, instead of the ethnonym *Ar'mân / Makedon-Ar'mân*.

Community, ethnicity, nation

Because we already have talked about the Aromanian community, we must bear in mind that the idea of *community*[8] means a group of people with similar interests, beliefs or rules of common life.

Ethnicity[9] and *people*[10] are defined as a group or community of people of the same origin, language and cultural traditions, and so are the Aromanians.

Unfortunately, though the Aromanians constitute a very old people, as Mr. Mantsu said in Tirana, they never had a country of their own, in order to move to a higher stage of development, the state of the *nation*.

---

[8] COMMUNITY Group of people with common interests, beliefs or rules of common life; all the inhabitants of a city, of a country etc. *DEX '98*
1. v. Community. 2. Ethnic community.

[9] ETHNICITY. Group of people of the same origin, language and cultural traditions.

[10] PEOPLE. 1. Historical form of human community, [...] whose members live in the same territory, speak the same language and have the same cultural tradition. [...]

EUGENE MATZOTA

One language, one nation, one state?

The idea of equivalence between language and *nation*[11], which leads on to the idea of a national *state*,[12] first appeared in the nineteenth century. As the term *nation*, *national* is not a word too old. It seems that the battle cry of '*Long live the nation!*' was heard for the first time during the Battle of Valmy in 1792.

The idea of a *national state* defined by the language spoken by the majority of those living in a certain area does not fit the model of ancient communities. The problem with so many individual agendas of the Balkan nations that also meant irredentism and imperialism, led to some irreversible 'effects notions of nationality', as pointed out by *Dr. Rodanthi Tzanelli*[13] in one of her lectures[14] in 2010. As the sources we rely on when we talk about antiquity are written generally in two languages that were spread then, the Latin and the Greek, it's hard to give credit to those documents.

---

[11] NATION Stable community of people, historically constituted as a state, appeared on the unity of language, territory, economic life and mental structure manifested in specific peculiarities of national culture and consciousness of the common origin and fate. [...] DEX '98
[12] STATE. Super structural institution, main instrument of political organization and administrative functionality through which carries social system and governs the relations between people; territory and population over which the organization exercises authority; country. [...] DEX '09
[13] Dr Rodanthi Tzanelli, Associate Professor in Cultural Sociology, University of Leeds
[14] *A History of the Balkans since the Nineteenth Century*, Imaging and Inventing the Balkans in Historiography, https://docs.google.com/viewer?a=v&pid=sites&srcid=ZGVmYXVsdGRvbWFpbnxyb2RhbnRoaXR6YW5lbGGxpc3RlYWNoaW5nfGd4OjE5NzNjOTk0MDc2MDJjZTE

Figure 1 - *Macedonia and Thessaly on the map of Greece, Historical Atlas by William R.Shepherd, New York, Henry Holt&Comp. 1911*

For the Greeks or the Romans, all the others were barbarians, which does not mean that they were necessarily inferior, just presented as barbarians. Therefore, their actions are clearly minimized even by historians who contend that they rely solely on facts, not on myths.

In some way, the Aromanians problem overlaps, unfortunately, the Macedonian issue. There are two schools of thought in Macedonia. Some scientists have sought to demonstrate that Macedonia is a sort of 'Slavic glory'. Others, on the contrary, tried to argue that Macedonia has always belonged to Greece.

In light of this situation, the problem lies in the fact that the Aromanians live mostly in the area where the ancient Macedonia used to be, an area containing Thessaly and Epirus in Roman times. Consequently, their problems are the same as those of the inhabitants living at that place.

On the other hand, we feel the necessity to highlight the fact that we are not aware of any evidence that the ancient inhabitants of Macedonia were somehow speaking the Slavic language. On the other hand, given the coming of the Slavs in the Balkan Peninsula, sometimes after the year 500AD, this might be impossible...

Pelasgians and Aromanians

Now, if we understand clearly that the Macedonians did not speak any Slavic language, we have to find out what kind of language they spoke from antiquity to the present day... Branislav Stefanoski - Al Dabija, who has written several books on the subject, believes that they were speaking Aromanian, a language unchanged for thousands of years.

> *'Aromanian language has not changed in the last 4000 years, [...] 'Latinization' is pure invention, [...] process actually went in the opposite direction [...] Romans are descendants the Aromanians, what is argued in detail in the writings of many ancient authors, Homer and Virgil, especially.*

> *The language arguments consist of the reading of those 24 inscriptions in the Aromanian language (Thracian-Illyrian) dating from the VII century BC until the second century A.D. on the territories of the Balkan Peninsula, Asia Minor and Kuwait.*
>
> *Among the most significant: 'Midas Epitaph from the VII century. BC', 'Kiolmenus Blocks', 'The Ring of Ezerovo ' [...] analysis of many words from Greek archaic Aromanian translation of the Homer's 'Iliad'.'*[15]

About the Pelasgians have written not only the great German historian *B. G. Niebuhr*[16], but also some renowned French historians, such as, for example, Jules Michelet[17]. The latter asks himself whether we should not explain the disappearance from the stage of history of such a great people and, also, the disgust manifested by the Greeks against the Pelasgians, by their contempt for those who had peaceful occupations, not warlike ones.

Niebuhr makes it clear, in a footnote containing a description of the relationship between the Pelasgians

---

[15] Stefanoski – Al Dabija, Branislav, *Scurtă istorie descriptivă despre originea makedon-armânilor (de la preistorie până la colonizarea Daciei)*, Ed. CNI Coresi, Bucureşti, 2011, p.14-15

[16] B. G. Niebuhr (1776 –1831), Danish-German statesman and historian. Germany's leading historian of Ancient Rome, founding father of modern scholarly historiography.

[17] Jules Michelet (1798 - 1874), French historian, supporter of the Romanian National Awakening movements..

and the ancient Greeks, that the Macedonia inhabitants were neither Illyrians, nor Thracians, but Pelasgians:

> *'In the earliest times, when the history of Greece is yet wrapped up for us in impenetrable mystery, the greater part of Italy, perhaps the whole of the eastern coast of the Adriatic, Epirus, Macedonia (The original inhabitants of Macedonia were neither Illyrians nor Thracians, bat Pelasgians. – Footnote 1*), the Southern coast of Thrace with the peninsulas of Macedonia, the islands of the Aegean as well as the coasts of Asia Minor as far as the Bosporus were inhabited by Pelasgians; that they were not barbarians is confirmed by the unanimous opinion of all the Greeks and may be seen from Homer; they inhabit a Pelasgian country hut their names are Greek.'[18]*

Now, let us see what *Pouqueville*, a diplomat and historian who has great knowledge about that area, has to say about the Pelasgians in *Voyages de la Grèce*, in the sixth book, *Dolopia or Anovlahia*:

> *'It does not enter into my topic to investigate whether domestic or Dolopians were descended from the Pelasgians. Homer,*

---

[18] NIEBUHR, B. G., *Lectures of The History of Rome From the Earliest Times to the Fall of the Western Empire*, London, 1850, Vol. I, The Pelasgians, p. 14

> *revealing their existence, teaches that*
> *they appear at the Siege of Troy, among*
> *Achilles' soldiers, son of Peleas, King of*
> *Tessalians, and they lived on the shores*
> *of the Gulf Pasagetic (Pelasgic or*
> *Pagasitikos, gr., E.M.).'[19]*

About the Pelasgians, Macedonia's ancient inhabitants and their descendants, also speaks *Felix Colson*, who is well acquainted with the realities of the mid-nineteenth century to the North of the Danube and beyond:

> 'What reveals us the language in respect to
> Pelasgian and Dacian nationality? What
> was the idiom spoken by the Vlachs?
> Philologists have considered it a
> takeover from the conquering Romans.
> There is only an assertion that seems less
> founded. Vlachs idiom is the Pelasgians
> idiom, it was made up of thirty centuries.
> It has been spoken in the Pindos
> Mountains, more than a hundred years
> before the conquest of the soldiers of
> Trajan.'[20]

---

[19] POUQUEVILLE, F.-C.-H.-L., *Voyage de la Grèce*, Tome second, Chez Firmin Didot, Paris, 1826, p.325

[20] Colson F., *Nationalité et régénération des paysans moldo-valaques*, E. Dentu, Paris, 1862, p.20-21

JUST A KIND OF GLOSSARY

*'All peoples from the Balkans
have been educated in the spirit
of megalomania.'*[21]

VLACH or AROMANIAN

This particular term, *Aromanian*, seems to be invented by Gustav Weigand when, in 1895, he published in Leipzig *Die Aromunen Ethnographisch-philologisch-historische Untersuchungen.*

Weigand explains in the foreword why he uses this term, *'generally spread among the people'*:

---

[21] Iorga, Nicolae, *CE ÎNSEAMNĂ POPOARE BALCANICE.* Conferință ținută la Ateneul Romîn în ziua de 13 Decembre 1915, Neamul Românesc, Vălenii-de-Munte, 1916, p. 21

After all, since one finally has to agree on a name, I have proposed the one that the people in all its fields gain for themselves, namely, 'Aromunen', which is the German rendition of 'Arămāni'.

It did not come to me to invent this name, or even to build it on the basis of Miklosich's 'Rumunen' form, as Gustav Meyer underlined this in his critique of Indo-European research.

For the Daco Rumanians one has a common form: Romanians, Miklosich's 'Rumunen' was superfluous, but for the 'Aromunen' there is no uniform form, so this is necessary.

So what was more natural than to leave their own name, but under the German form; the Romanians may write Arămâni or Armâni, it is for them, but not for Germans, because the sounds ă and î are unknown to them.[22]

For those who want to find out more from Weigand explanation, a longer excerpt from his book could be found in the ANNEXES.

Now, we use here *seems to be invented by* because, in German, the term *Romanian* is translated by *Die Rumänen*, not *Die Romunen*...

Therefore, *Die Aromunen* is just similar to the correct version *Die Arumänen*, and this is only one of the

---

[22] Weigand, Gustav, *Die Aromunen. Ethnographisch-philologisch-historische Untersuchungen*, Leipzig, 1895, p. VI-VII

slightly distorted things, but so important, that have a history of misleading longer than a century. And we will stop right here, just mentioning this, because this book has much more to say than to discuss about a distorted misleading translation, no matter how important this translation might be. This is a problem for the linguists to solve, not for us.

That is why, for the events prior to the nineteenth hundred, I generally used the term Vlachs, and, from that moment on, Aromanians, as they name themselves, after all (in fact, *Arm'ns*).

Solely in order to simplify things I have made this decision, as there still exist so many variants, inappropriate for the amateur reader, who simply desires to find out more information on the topic. That is why, the information, as it is revealed from the sources, is generously presented in the notes and in the appendices.

In this way, I believe that I will shed light on so much more problems for those who desire to reach the sources, for those who would like to have a better understanding about the Aromanians in the communities where they lived, with those names given by their neighbors, by the others who met them during their travels, etc.

VLACH or MACEDONIAN

The difference between *Vlachs* and *Macedonians*, or *Makedon-Aromâns*, which is the term used by *Branislav Stefanoski – Al Dabija* his well-documented papers, is made, for the first time in history, by the chronicler *Barensis*.

Barensis, while retelling the Byzantines' expedition in Sicily in 1027, says that in it took part both Vlachs and Macedonians (*Vlachorum, Macedonum*, in Latin in the original text)[23]!

In this case, for instance, as Stefanoski-Al Dabija himself draws our attention, even though the chronicler Barensis uses the term *Vlachorum*, which makes us think about *Vlachs*, I can only understand and refer to *Valachs* in this context, since a distinction between the two is made in the same sentence, meaning the difference between *Vlachorum* and *Macedonum*.

Another hypothesis might be, for instance, the identification of the Aromanians of Thessaly as Vlachs, to distinguish between them and those from Macedonia.

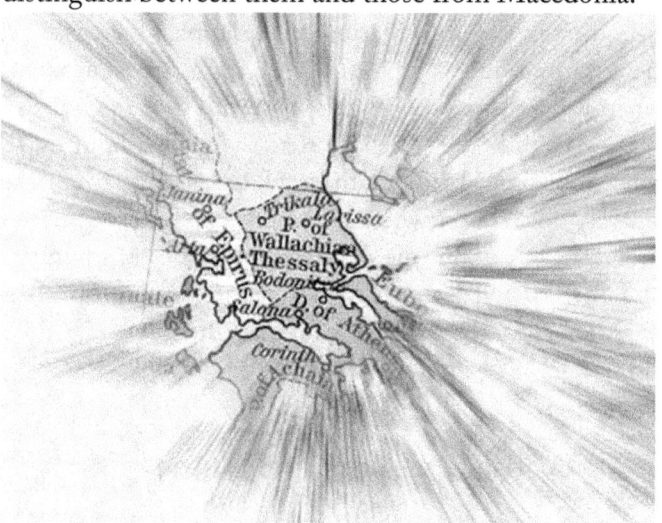

Figure 2 – *Great Vlachia, next to Epirus*

---

[23] "Burgarorum, Vlahorum, Macedonum, aliarumque ut caperet Siciliam". - Pertz, Georg Heinrich, Monumenta Germaniae Historica. Anales Barensis, tom V, p. 53

For whoever would notice on any map the place of the ancient Macedonia and the ancient Thessaly, called *Megalovlahia*, that is to say *Great Vlachia (Vlaquie, Vlakie, Țara Vlahilor)*, things could be easily clarified.

We only want to point to some problems that have been superficially searched or deliberately distorted in the histories of the peoples of the Balkans.

VLACH or VALACH (WALACH)

We have tried to keep up with the consistency of the message when the historical documents speak alternately about *Vlachs* or *Valachs*, actually the reference is the same – the same *Vlachs*.

> '*On the grounds of the historical sources and of the stages chronologically run in succession by the settlement and social life of the Aromanians, it results that the name of Vlach makes its appearance on the coming of the Slaves in the East, so that the word Vlach can't be but of Slavic origin.*

Figure 3 - *The Byzantine Empire, 1265. Historical Atlas by William R. Shepherd, New York, Henry Holt&Comp. 1911*

By the word Vlach, the Slaves were indicating the people of Romance origin, in the same manner the Germans referred to the people from their regions who were of Romance origin, like the Vlachs, as Welch and Wallon.

The name Vlach has remained intact to this day, and we can find it in all the languages. The Serbians say Vlach; the Bulgarians, Vlach; the Poles, Woloh; the Czechs, Valah; the Russians, Woloh; the Turks, Vlach; the Hungarians, Oloh; the Italians, Vlacho; the French, Valaque; the Germans, Walach and the Greeks say Vlachos.

Figure 4 - Macedonia and Thessaly on the map of Greece,
Historical Atlas by William R. Shepherd, New York, Henry
Holt&Comp. 1911

**The word Vlachos is present both in the poetry
of Greek modern poets and in the Greek
folk songs, especially in the outlaws'
ones.'[24]**

Thessaly is also known by the name of those inhabiting it, the Vlachs, according to the great French byzantinologist G. Schlumbergerin in *Les Principautes Franques du Levant*, page 61: *"Thessaly was then more familiar to us under the name of Vlaquie, Vlakie, Terre des*

---

[24] Diamandi-Aminceanu, Vasile, *Românii din peninsula balcanică*, București, 1938, p. 29

Vlaques, or Megalovlaque", apud *Vasile Diamandi-Aminceanu.*

The origin of the word *Vlach*, with all its variants, seems to be related to a confederation of Celtic tribes, called *the Volcae*, spreading simultaneously in the South of France, but also in Moravia, to reach as far as Galatia, in Asia Minor.

> '*Volcae*, in ancient Gaul, [was] a tribe divided into two parts: testosages, on the superior valley of the river Garonne around Tolosa [Toulouse], and arecomici, on the right banks of the river Rhone, having their center in Nemasus [Nimes] [...]
>
> Testosages participated in the Celtic invasion of Greece and of Asia Minor, during the Ist century B.C. [...] '[25]

The *Volcae* tribes held an important position in Moravia, being among those who controlled the commercial routes between the Germanic territories and the Mediterranean Sea. The Germans name them with a generic term proper for and referring to both the Celts and the Romans: *Walha*, the word used for any of the ex-inhabitants of the Roman provinces, being borrowed by the Slaves to designate the Vlachs, and comparable to the English term *Welsh* or with the Dutch one *Walloons*.

---

[25] *Encyclopedia Britannica,*
http://www.britannica.com/EBchecked/topic/1574478/Volcae

In the long run, *Walha* is used more and more frequently in the Roman provinces of Western Europe, under the form of *Walch* or *Walach*. Starting with the tenth century, the Southern tribes of Slaves use *Vlach* to refer to a foreigner or somebody who is not Slave, but speaking a Romance language.

We should not forget about the Moravian *Valachs* (*valaši*, in Czech language). In a DNA study[26] from 2011 special importance is laid on their explicit difference from their surrounding neighbors and the closeness to some isolated populations in the Balkans, such as the Aromanians.

[...] The Valachs (or Wallachs/Vlachs as they are sometimes called) are one of the most distinct ethnographic and cultural subpopulations of Central Europe.

Today, they can be found not only in the Czech Republic – in its eastern border mountain ranges and highlands (Beskydy in Moravia) – but also in South-Southeast Poland and several parts of Slovakia (far western, Northern, and central region).

Originally, this group spread from the Maramures region of Romania, roughly following the Carpathian Mountain range. The arrival of the Valachs to the area of today's Czech Republic took

---

[26] Ehler et al, *Y-chromosomal diversity of the ValachsY-chromosomal diversity of the Valachs from the Czech Republic: model for isolated population in Central Europe*, CMJ, 2011, http://cmj.hr/default.aspx?ID=11764

*place at the very end of the 15th or
beginning of the sixteenth century (12).*

*The migration was not spontaneous, but
rather encouraged and subsidized by the
local nobility, and it lasted at least until
the end of the eighteenth century, with
immigrants supposedly coming not only
from Romania, but also from Ukraine,
Poland, and Slovakia.* [27]

Since their coming here, in the fourteenth
century, they have been called either *Vlach* or *Valach*,
Volach and *Vlakh*, to result into the today's term: *Valaši*. In
the old Czech language, the word *Valassko* means *Italy*,

Figure 5 - Vlachs populated area in Moravia, Valašsko
region, in the Northwest of the Carpathians, and in the
nearby area

---

[27] Ibid.

and the term *Valach* referred to an Italian, which can be also certified in other languages.

The place of their origin is still a mystery, but their vocabulary comprises words characteristic for the shepherd's occupation and familiar for Aromanians or Romanians.

May they be the descendants of those who were inhabiting the Pannonian plans before the Hungarians arrived, as an anonymous monk writes in a chronicle entitled *Descriptio Europae Orientalis*, calling them *Romanorum pastores*?

Their culture was mainly destroyed at the end of the 30 Years War, after *The Peace of Westphalia*, in 1648, a war in which the *valaši* were directly involved.

ABELA, not AVDELLA

Concerning the names of the localities of the Balkan Peninsula, I have used the way they were written in the original, trustworthy documents concerning their precision at the time of the end of the nineteenth century and the beginning of the twentieth century.

In *Lumina, a popular magazine of the Romanians of the Ottoman Empire 'Publication of the educational and religious staff of Turkey*[28], of October 1905, with the editorship at the *Romanian High School* of Bitolia, we can find some statistics of the students who attended that school, mentioning the names of the localities they came from.

---

[28] Lumina, revistă populară a românilor din Imperiul Otoman, „Publicație a corpului didactic și bisericesc din Turcia"

We do believe that those teachers, finding themselves at that particular moment in time there, among the Aromanians, and not in Romania, for example, knew better than anyone else the names of those localities in the Aromanian language.

Consequently, I will write *Abela*, as our teachers were saying, and not Abdela, Avdela or Avdella. The last version mentioned here, *Avdella*, is provided by *Wikipedia*, which is not a scientifically reliable source.

We have to remind you that this online encyclopedia, Wikipedia, is an honorable information source, but, being supported by volunteers, we cannot credit them for their objectivity. Moreover, Wikipedia warns us that it is not responsible for the accuracy of its contents.[29]

Likewise, I will write Cruşova, not Krushevo. That is why I preferred this type of orthography before any other type of writing taken over from a dubious text in English or, even more confusing, in Greek. We will use *Bitolia*, *Bitola*. Even less could I use, illogically, instead of Bitolia, the Turkish-originated variants *Monastir* or *Manastir*.

---

[29] „[...] the content of articles and other projects is for informational purposes only and does not constitute professional advice." https://wikimediafoundation.org/wiki/Terms_of_Use

Figure 6 – *Lumina, Popular Magazine of the Romanians of the Ottoman Empire – COVER*

Numărul absolvenților după localitate

No. de ordine	LOCALITATEA	Profesori	Studenți	Revizori	Institutori	Pedagogi	Doctori	Farmaciști	Avocați	Ingineri	Funcționari	Ofiteri	Fotografi	Comercianți	Agricultori	Fără ocup.	Morți	Total
1	Abela	1	9	1	2	2	2	—	—	—	—	—	1	8	1	1	4	32
2	Băiasa	—	3	—	1	—	—	—	—	—	—	—	—	3	—	—	—	6
3	Berat	—	—	—	—	—	—	—	—	—	—	—	—	1	—	—	—	1
4	Belcamen	—	1	—	—	—	—	—	—	—	—	—	—	—	—	—	—	1
5	Brusa	—	1	—	—	—	—	—	—	—	—	—	—	—	—	—	—	1
6	Breaza	—	1	—	—	—	—	—	—	—	—	—	—	—	—	—	—	1
7	Crușova	—	4	—	—	—	—	1	1	—	2	—	—	—	—	—	—	8
8	Clisura	—	2	—	5	—	1	—	1	1	1	—	—	—	—	—	—	11
9	Duratzo	—	—	—	1	—	—	—	—	—	—	—	—	—	—	—	—	1
10	Elbasan	—	—	—	1	—	1	—	—	—	—	—	—	—	—	—	—	2
11	Ferica	—	—	—	1	—	—	—	—	—	—	—	—	—	—	—	—	1
12	Florina	—	—	—	—	—	—	—	—	—	—	—	—	—	—	1	—	1
13	Furca	—	—	—	1	—	—	—	—	—	—	—	—	—	1	—	—	2
14	Gopeși	2	2	—	1	—	1	—	—	—	1	—	—	—	—	—	1	8
15	Huma	—	—	—	1	—	—	—	—	—	—	—	—	—	—	—	—	1
16	Hrupișta	—	2	—	1	—	1	—	—	—	1	—	—	—	—	—	—	5
17	Magarova	—	2	—	—	—	—	—	—	—	1	—	—	—	—	—	—	3
18	Aminciu	—	1	—	1	—	—	—	—	—	—	1	—	—	—	—	—	3
19	Caterina	—	—	—	1	—	—	—	—	—	—	—	—	—	—	—	—	1
20	Bitolia	3	2	—	—	—	—	1	—	—	2	—	—	3	—	1	—	12
21	Mulovișta	—	2	—	—	—	—	—	1	—	—	—	—	—	—	—	—	3
22	Moscopole	—	1	—	—	—	—	—	—	—	—	—	—	—	—	—	—	1
23	Meglen	1	—	—	3	—	—	—	—	—	—	—	—	—	—	—	—	4
24	Neveasta	1	4	—	1	—	—	—	—	—	—	—	—	—	—	1	—	7
25	Nijopole	2	1	—	3	—	—	—	—	—	—	—	—	1	—	—	3	10
26	Ohrida	1	1	—	3	—	1	—	—	—	—	—	—	—	1	1	—	8
27	Perlepe	—	4	—	2	—	2	—	—	—	—	—	—	—	—	—	—	8
28	Perivole	2	2	—	7	—	—	1	—	—	1	—	—	4	—	—	1	18
29	Pleasa	—	—	—	1	—	—	—	—	—	—	—	—	—	—	—	—	1
30	Papadia	—	—	—	1	—	—	—	—	—	—	—	—	—	—	—	—	1
31	Resna	—	—	—	1	—	—	—	—	—	—	—	—	—	—	—	—	1
32	Samarina	1	2	—	4	—	—	—	1	—	2	—	—	4	1	—	—	15
33	Smixi	—	—	—	1	—	—	—	—	—	—	—	—	—	—	—	—	1
34	Selea	—	1	—	—	—	—	—	—	—	1	—	—	—	—	—	—	2
35	Tărnova	—	—	—	2	—	1	—	1	—	—	—	—	—	—	—	—	4
36	Tirana	—	—	—	—	—	—	—	—	—	—	—	—	—	1	—	—	1
37	Turia	—	2	—	—	—	—	—	—	—	—	—	—	2	—	—	—	4
38	Veles	—	1	—	—	—	—	—	—	—	—	—	—	—	—	—	—	1
39	Veria	5	6	—	6	1	—	—	1	—	2	—	—	1	—	2	2	27
		19	56	2	52	4	11	3	5	1	14	1	1	29	3	7	13	218

Figure 7 - Fragment from number 10 festival, October 1905 of *Lumina, a popular magazine of the Romanians of the Ottoman Empire 'Publication of the educational and religious staff of Turkey''*, dedicated to the 25th anniversary of the foundation of the Romanian High School of Bitolia, with a list of localities, with the names used at that time by their inhabitants.

A HISTORY OF HISTORIES

Why necessarily a history of histories?

> *'There is in all Balkans no race so mysterious and individual as the Vlachs.* [30]

This title means that I am trying to create a sort of honest summary of the most significant theories regarding the Aromanian issue. By *significant*, I understand nothing more than the fact that they have reached my ears.

I have reviewed most of the theories launched in the second half of the nineteenth century, when the new Romanian national state became aware of the Aromanians' issue. This is a time when the Romanians began to worry about their brothers in the South. Several Romanian schools appear in the Balkans, in a manner which might now raise the question why Romania was trying so hard to impose the Romanian language, instead of maintaining the Aromanian one, as the great Romanian historian *Nicolae Iorga* said in 1928.

---

[30] Brailsford, Henry N., *Macedonia: its Races and their Future*, London, 1903, p. 175-176

We do not want to discuss the way the *Aromanian issue* was treated in those troubled times. Conceivably, now we just see things differently...

Because I do not want to invent anything, nor do I think that I can find the right answer to questions that nobody could give a clear answer yet. There are plenty of theories, and so are plenty of answers...

I have found, in the end, that an honorable solution would be to honestly try to make available more reliable information, including DNA studies, to those who want to understand more than some partisan position papers, either intentional or not, related to different interests and different periods of time.

This book, in order to be closer to the Truth, contains many quotations, specifically in order to provide the reader with what I could gather from sources which I considered to be the most honorable. Anyway, as honest and especially credible as extremely ancient chronicles might be. What I probably fetched back are some theories coming from experts that cannot be accused of whatever kind of partisanships.

This is a book of questions and certainties based on historical data.

The answer might be different, according to the truth each one of us claim to detain...

In the end, anybody has an answer, depending on his or her own truth...

Mythistory, myth and/or history?

This term, *Mythistory*, was coined by the great historian William H. McNeill[31] in 1985 and appears in the very title of his speech[32] at *The American Historical Association*, as president.

Right from the start, McNeill highlights the fine line that rather separates than approaches these two terms, namely *myth* and *history*:

> *'Myth and history are close kin inasmuch as both explain how things got to be the way they are by telling some sort of story. But our common parlance reckons myth to be false while history is, or aspires to be, true. Accordingly, a historian who rejects someone else's conclusions, calls them mythical, while claiming that his own views are true. But what seems true to one historian will seem false to another, so one historian's truth becomes another's myth, even at the moment of utterance.'[33]*

This method of mixing scientific data with mythological data that interprets real historical facts may be an excellent model in the Balkans. Here, one might

---

[31] William H. McNeill, the author of "The Pursuit of Power, The Rise of the West" (1963) and "Mythistory and Other Essays" (1986), published de *University of Chicago Press*.

[32] Mythistory, or Truth, Myth, History, and Historians

[33] McNeill, William H., *Mythistory, or Truth, Myth, History, and Historians*, The American Historical Review, Vol. 91, No. 1, Supplement to Volume 91 (Feb., 1986), p.1

find several historical truths, because the Balkan peoples have specific myths that are always overlapping geostrategic needs.

Balkan historians almost always molded their vision on the aspirations for bigger and for more of the peoples they come from and for whom they wrote. Thus, history can sometimes be a myth for somebody else, and our heroes may as easily be your traitors. And vice versa...

And all depends on the perspective from which me and you are looking at things...

The Aromanians have lived since we know from scientifically verifiable sources on a vast territory that overlapped somewhat over the Eastern Roman Empire, then over the Ottoman one.

In any case, this territory goes far beyond the contemporary borders of the Balkan states. That must be the reason why the Aromanians have never ever entered this game of historical myths associated with these Balkan nations.

The historian, patriot or traitor?

The Aromanians did not have the time to write their history, they had their daily work, so we cannot talk about the Aromanians' histories or chronicles written about them. This applies until the nineteenth hundred, when the emergence of the national sentiment causes, particularly in Romania, a need for the search of common roots with their brothers South of the Danube.

Each historian writes about what he has witnessed, read or heard about something, from his own perspective. If you want to make out some clutter in

Figure 8 - *Pastor and boys from Lânga,*
Weigand, Die Aromunen Erster Band. p. 63
Digitized by Google

history where the Aromanians are also summoned, you must always bear in mind the vision of different versions of the historical truth, when it comes to the truth of the others...

Until the twentieth century, several people who were in one way or another in contact with the Aromanians, apart from the Romanian brothers across the Danube, spoke about them.

Much quoted in the works devoted to the Aromanian Question is *François Pouqueville*[34]. General

---

[34] *François Charles Hugues Laurent Pouqueville* (1770 – 1838), French diplomat, writer, historian etc., French Institute member.

Consul of France in Ianina, he experienced the Balkans life in those times, writing a five-volume *Voyage from Greece*, published in Paris between 1820 and 1822.

We will speak afterwards about what almost all major travelers that visited the Ottoman Empire found there. Of course, for those in the world who met rigorous Protestantism, the Balkans are fascinating.

It was a world that knew no Renaissance, and so interesting, being a glamorous mystery, that Mozart has an opera that takes place in a seraglio.[35]

If we interpret the myth as a retelling of the historical realities that could be real, we understand the difference between history and myth. To be short, history tells us what happened and the myth explains the meaning.

W. H. McNeill clarifies the historian's dilemma of choosing between myth and historical fact through a human approach to the problem, which excuses reluctance for anything that is not scientific.

> '*A century and more ago, when history was first established as an academic discipline, our predecessors [...] believed they had a remedy. Scientific source criticism would get the facts straight, whereupon a conscientious and careful historian needed only to arrange the facts into a readable narrative to produce genuine scientific history. And science, of course, like the stars above, was true and eternal, as Newton and Laplace had demonstrated to the*

---

[35] *Die Entführung aus dem Serail*, by W.A. Mozart, 1782, Vienna.

> *satisfaction of all reasonable persons*
> *everywhere.* [36]

Comparing the evolution of the historians' approach and that of physics, we might say that purely scientific history is physics before quantum physics, a blasphemy for the followers of Thucydides.

Joseph Mali[37] begins his book *'Mythistory: The Making of a Modern Historiography'* with an excitingly captioned and very conclusive title: *Where Terms Begin: Myth, History, Mythistory*:

> *'Ever since Herodotus declared in Histories*
> *that to preserve the memories of the*
> *great achievements of the Greeks and*
> *other nations, he would count on their*
> *own stories, historians have debated*
> *whether and how they should deal with*
> *myth.'*

As Mali says so well, behold, from Herodotus onwards, historians had to choose between the path proposed by him, which is to include in their work, their own historic stories and myths of the peoples about whom they wrote, and Thucydides' way, who says that Herodotus' method was not a scientific one.

---

[36] McNeill, William H., *Mythistory, or Truth, Myth, History, and Historians*, The American Historical Review, Vol. 91, No. 1, Supp. to Volume 91 (Feb., 1986), p.1

[37] Joseph Mali, Dep. of History Tel-Aviv University. "The Rehabilitation of Myth: Vico's New Science" (1992) and „Mythistory: The Making of a Modern Historiography" (2003).

Disputes between the two trends were conducted throughout the centuries between *Montesquieu* and *Machiavelli*, between *Mommsen* and *Niebuhr*, to cite just two of the most known pairs.

Modern historians, particularly those of recent decades, tend to merge these two views. Mali proposes a reconciliation of the two directions, pleading for a historiography that recognizes the importance of myth in the construction of identities, in a search for historical reality.

The history of myth or the myth of history?

Felipe Fernandez-Armesto, one of the important historians nowadays, owes his books' success to the extraordinary manner of narrating tales. He is known for his view of world history written from a different position, different from ours, of the earthlings limited even by our limitations, by what we believe at some point to be right or not.

In his book '*1492: The Year Our World Began* ', Fernandez-Armesto begins with '*This world is small*'. And so it is, the world of each of us is really a small world. In a bigger world, that of the Balkans, there are more small worlds, each with its own history, its own values and truths that can only be true for the world of which they are created.

To grasp the intricate issues of the Balkan histories, we must understand what Fernandez-Armesto meant when he talks over the question of truth in philosophical terms.

He distinguishes between my truth and your truth, our truth and yours and these truths can coexist in the search for absolute truth.

'POLITICALLY CORRECTNESS' AND HISTORY

That lie called 'politically correct'

The slogan of the Theosophical Society says there is virtually nothing higher than truth. Therefore, I, as a Theosophist, I might understand Fernandez-Armesto when he says that we should not give up seeking the truth while fighting against lies, primarily, but also against the false modesty of not telling the truth just because it could touch someone's feelings, a false attitude tolerance, a lie, actually something *politically correct*.

> 'To adhere to the search for truth does not mean the abandonment of skepticism.
> [...] I tried to show that different cultures, at different times, have favored, on balance, different techniques for

telling truth from falsehood, and therefore may be said to have had, to that extent, different concepts of truth. In that sense, the predictions of relativism have proved valid. And it is important to remember that truth is elusive. It takes hard work, discipline, and time to approach it. Although the truth is out there we shall not grasp it quickly or easily embrace it whole [...].[38]

In those insignificant worlds of the Ottoman Empire, with their small and narrow ideals that tend to find their identity in order to become independent nations, myths and histories differ. Especially after the revolutions of 1848, the myth of the heroic history tends to turn into real history. If you can talk somewhere about intertwining histories of neighboring peoples with histories of other peoples, the Balkans appear to be the most appropriate place.

In her book *Albanian Identities: Myth and History* (2002), *Stephanie Schwandner-Sievers*[39] speaks about mythistory effects on the Aromanians of Moscopole. In the Albanian history books, *Sali Butka* is described as a great patriot of Albania during the Balkan Wars and then World War I. Among his countless acts of heroism is the murder of traitors in Moscopole.

For the Aromanians in Moscopole, who know another truth, those traitors are nothing but heroes.

---

[38] Interview, *Pulse Berlin*, http://www.pulse-berlin.com/index0985.html?id=146

[39] Dr. Stephanie Schwandner-Sievers, social anthropologist, expert in Albania and Kosovo, University of Bournemouth.

> *'There was an aspect of subgroup self-control in the fact that these conflicting interpretations of history were not put on a public stage during times of repeated national crises in the 1990s. Despite their new collective rights, Aromanians learnt a bitter lesson after press reports made them national scapegoats for political failures [...] Luckily, the Aromanian self-image included reference to ideals such as education, tolerance, mobility, flexibility and peacefulness—which may have guided their conflict-avoidance strategies in times of crisis.'* [40]

This is as if Hitler had won the war and all those heroes of the French Resistance, for example, would become terrorists trying to destabilize the Great Reich.

For some, terrorists, for others, heroes. It depends on whom writes the history.

And, we must not forget, history is written by those who win, not by the losers.

A world only for the Aromanians

Those borders so much desired by Balkan nationalists, by those dreaming of a Great Bulgaria or a Greater Serbia, from case to case, are nothing but some setbacks for the Aromanian. With a particular way of living, with specific transhumance and great talent to

---

[40] *Stephanie Schwandner-Sievers, Bernd J. Fischer*, Albanian Identities: Myth and History, C. Hurst Se Co. (Publishers) Ltd, 2002, p.16

trade, an Aromanian cannot be closed between any narrow national boundaries. Much larger space, as the Ottoman Empire or the Habsburg one in the past, and the European Union today, are perfect for Aromanians, because the generous extent of the Ottoman Empire and then the one of the Habsburg Empire represented a guarantee of freedom of movement that fitted like a glove.

This is probably the essential difference between the Jews and the Aromanians, despite many similarities in their social position and appetite for trade: the lack of a Zionist movement type. Unlike the Jews, the Aromanians do not have a movement for national emancipation and, more than this, that motherland of the Jews.

In this context, I believe that is not possible to hear any Aromanian saying: *We, Aromanians, we are the rulers here, on this piece of land!* Or, moreover: *this land is ours!*

This would be the story about the Balkans. This piece of land that is now yours, was mine...

> '*The same truths look different when viewed from different viewpoints. Truth, as I am always telling my students, is like a nymph glimpsed bathing between leaves. The more you shift perspective, the more is revealed.*'[41]

During several discussions, I have had with people from the former Tito's Yugoslavia, I have often

---

[41] Felipe Fernandez-Armesto, interview in *Pulse Berlin*, http://www.pulse-berlin.com/index0985.html?id=146

heard cases of radical change in the attitude towards colleagues or friends in the early 1990s. It was a time when communism constraints and cohesion disappeared over the night. Suddenly, the world has been divided into two: our world and your world!

People who used to be friends and grow up together, have now become fierce enemies, in the name of the heroic stories that everyone knew from his ancestors and that he thought were just about the past.

Unfortunately, it seems that in this area of the world the land requires blood and nobody forgets whose land it was before, especially if you have something to ask from his neighbor.

And again, unfortunately, there is always someone to reclaim something.

Territorially, the Aromanians do not claim anything, ever.

In a united Europe, including the Balkans as a whole, the Aromanians should feel at home, as always: honorable prominent citizens of Europe.

'Politically correct' Latinization

In order to have a better understanding of what the Aromanian Question might imply, we have to see what *Latinization* means for both Aromanians and Romanians.

By definition, *Latinization*, mostly known as *Romanization,* means the adoption of the Latin language and Roman culture in the area of influence of the Roman Empire. Comparing the map of the *'roman coloniae'* in the Roman Empire with the one of Romance speaking

Europe, we can see a big Latinized island in the North of Danube, namely nowadays Romania and Moldova, and some small ones in the Balkans, in the South.

The Romanization in that area of the Roman Empire has produced the Eastern Romance languages, spoken mainly by Romanians and Aromanians. When the idea of *nation* entered the political scene from the eighteenth century onwards, the historiography of the Habsburg and Russian Empires tried to deny the Romanization of the Romanian people, placing this process exclusively South of the Danube.

Thus, Romanians would be some kind of *'late immigrants'* in the territories they claimed, namely nowadays Romania and Moldova. On the other hand, the historiography of Balkan states, as Serbia, Bulgaria or Greece, regions with Aromanian minorities, would say that they are *'late immigrants'* arrived from the North of the Danube.

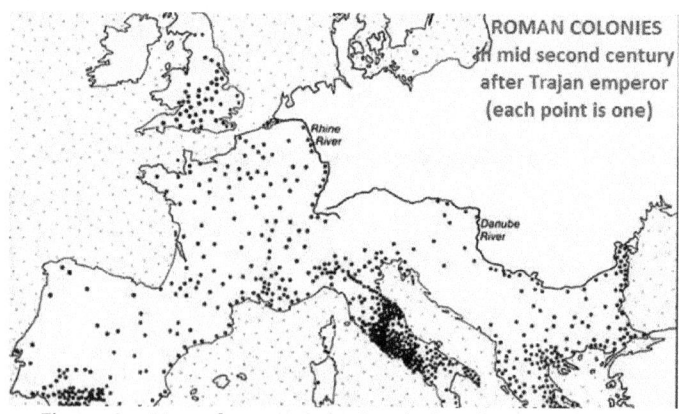

Figure 9 - *Map of roman coloniae during the second century*
Roman coloniae (*Wikipedia Commons*)

Now, could we believe that theory of the Aromanians' origin from Dacia, which would have been Latinized during the time of the Roman conquests?

We have to keep in mind a historical fact: the colonization of Macedonia by the Romans occurs long before the colonization of Dacia.

As a Romanian, I have been taught since childhood that a small part of Dacia became a Roman province (*Dacia Traiana* or *Dacia Felix*) in 106 A.D., after the Dacians had beaten the Romans often, but not often enough, as we were able to see. We also know that the Dacians were meddling in the politics of the Roman

Romance-speaking Europe
Romance language sole official and used by the majority
Romance language co-official and used by the majority
Significant non-Romance language usage or bilingual
Significant unofficial / historical Romance language usage
Roman Empire Latin limit

Figure 10 - *Romance-speaking Europe*
(*Wikipedia Commons*)

Empire, which would be the equivalent of meddling with Brussels nowadays.

Well, in order to do that, you have to rely on information; you have to be sufficiently influential and knowledgeable in making such inferences. It's very hard to believe when it comes to a barbaric people, from which it appears that we have nothing left, not even the language!

These things cause many to believe that the language of the Dacians is actually a vulgar language spoken by the people, not bureaucratic Latin. Of course, the influence of those old theories about the Latin origin of the Romanian people is strong, and it is hard to overthrow all that scaffolding built over the time.

If you dare today not to think that the Dacian women were hurling themselves together with their children into the arms of the Romans, and they had not even taught them their mother tongue, but that of the foreigner's, this is blasphemy! Specifically, as Felipe Fernandez-Armesto said, it is not 'politically correct', that is against the *System*.

And because we are mostly ruled by common sense, this makes you behave in such a manner as not to disturb the System, because you might be somehow not *'politically correct'*.

How could you say you are against those teachings shared by many generations of such honorable teachers and professors? Particularly since some of them were even your teachers!!!

So many scientists have fed this theory for more than 250 years and have earned the bread this way. Can you take the bread from their mouth?

It's not 'politically correct'. Better, just keep silent...

It's understandable, after all.

I might even say: *You know, this was 'politically correct' three hundred years ago...*

Anyhow, if we count the years since the formation of the Roman province of Macedonia, exactly 146 B.C., we see that Latinization could have begun in Macedonia more than 250 years earlier. But, what a miracle! Macedonia is not a Latin country!

Then, who was the first, the Vlach of Dacia or the one in Macedonia? Moreover, what influence could have had those so-called Vlachs coming from Dacia across the Danube, in the South, where there had already been Romanized elements for 230 years?

Difficult to answer...

Then, where are the Aromanians coming from?

The Aromanians' history, or *the Aromanian Question*, as the prestigious author of *Die Aromunische Frage*, Max Peyfuss, says, seems hard to understand and even harder to write about it.

Because we cannot make from all that I have heard, read or have written themselves, or from what we have witnessed with our eyes, more than just another history, we will label this part as a *History of Histories*.

Did the Aromanians come across the Danube, chased by the invasions of the migratory peoples? A hypothesis that, corroborated with specific transhumance pastoralism, could explain their presence in Moravia, as in

Southern Poland, but nobody knows where they were coming from or whom they were.

In 1933, V. Diamandi-Aminceanu makes an interesting recap of the situation of the Aromanian origin in his book *Romanians from the Balkan Peninsula (Românii din peninsula balcanică*[42]:

> **'The opinions of modern writers on the origin of the Aromanians from the Balkan Peninsula are different. Some[3] argue that Aromanians are remnants of the Roman settlers from the times of the Senate and of Empire[4] Others say that the Aromanians come from the settlers from Dacia. Others[5] again argue that Aromanians are a mixture of Moesians, Thracians, Bessi and Romanized Illyrians. Others[6] write that Aromanians are a mixture of Roman settlers from Romanized Dacia Traiana with Moesians.'[43]**

Apud V. Diamandi-Aminceanu, only one of those who treated the question of the Aromanians' origin, namely *Bidermann*, says in *Die Romanen in ihrer Verbreitung in Oesterreich* that they were Romanized Ligurians and Celts.

---

[42] Diamandi-Aminceanu, Vasile, *Românii din peninsula balcanică*, București, 1938, p. 12

[43] See ANNEXES, References to known authors and works more or less to be related to the question of the Aromanians origin...

Figure 11 - *Albanian monks, Aromanian shepherd and
merchant from Moscopole,
Weigand, Die Aromunen Erster Band, p. 104
Digitized by Google*

A much-quoted author when it comes to
Aromanians is *François Pouqueville*. Pouqueville describes
not only the Vlachs' life, but he also tries to find answers
about their origins, their name, to clarify the either
deliberate or not confusion between Bulgarians and
Vlachs, especially when it comes to events that had
marked the history of the Byzantine Empire.

Those who want to read a part of this
fundamental book for understanding the Aromanian
issue, some fragments could be found in the original
French language in the annexes. We preferred to leave it

as it is, not to distort Pouqueville's way of thinking, because *traduttore, traditore.*[44]

Gustav Weigand, the inventor of the term *Aromanian*, used different other words, as 'Olympo-Walachen' in 1888 in *Die Sprache der Olympo-Walachen*, 'Vlacho-Meglen' in 1892 in *Vlacho-Meglen. Eine ethnographisch-philologische Untersuchung.* Finally, in 1895, he coined the term 'Aromanian' in *Die Aromunen / Ethnographisch-Philologisch-Historische Untersuchungen.*

In *Short descriptive history about the origin of the Arm'n-Macedonians (From pre-history to the colonization of Dacia)*, Branislav Stefanoski – Al Dabija says:

> **'Arm'n Macedonians are direct successors of the Ancient Macedonians (Pelasgians-Thracians-Illyrians) and, in their everyday use, has almost completely saved the Ancient Arm'n-Macedonian (Pelasgian-Thracian-Illiric) language.'[45]**

Going beyond the theory of Latinization, Stefanoski - Al Dabija finds obvious similarities between Aromanians from the south of the Danube and the inhabitants of Maramures and Moldova, both in language and in port dress. For instance, specific Aromanian *fustanella* does not seem anywhere else north of the Danube.

---

[44] *traduttore, traditore*, IT.: translator, traitor, EN.
[45] Stefanoski – Al Dabija, Branislav, *Scurtă istorie descriptivă despre originea makedon-armânilor (de la preistorie până la colonizarea Daciei)*, Ed. CNI Coresi, Bucureşti, 2011, p.103

> '[...] *the Arm'n-Macedonian language, in the continuity of 4,000 years, has not undergone significant changes, so that the pre-historical, the ancient Arm'nian language is completely part of the present Arm'n-Macedonian language.*
>
> *I have come to this after reading many ancient scriptures written on stone blocks and relics belonging to the Arm'nian (Pelasgian-Thracian-Illyrian) tribes as well as analyzing the given translation of the Homer's' work 'Iliad' to archaic Greek language.'*[46]

When things are so complicated, the explanations of the Aromanians origin can be very different. *Spiru Lambru*, Aromanian born in Thessaly, Greece's prime minister, says in *Istoria tis Ellados*, Athens, 1898, apud V. Diamandi-Aminceanu:

> '*Vlachs Romanized descendants of the ancient Gauls subjected by Traian, who lived first besides Istros and Sau leading a nomadic life, dismount during the winter in Thessaly and Epirus, climbing from April to September with their cattle to the highest and wooded mountains of Macedonia and Bulgaria.'*[47]

---

[46] *Ibid.*, p.6-7

[47] Lambru, Spiru, *Istoria tis Ellados* t. V. cap, XXV p, 434 Athens, 1898

Figure 12 - *Aromanian family from Pleasa (right an Albanian).*
Weigand, Die Aromunen Erster Band, p. 113
Digitized by Google

We must draw the attention to the concept of *nomadic* here when talking about the Aromanians seen as shepherds. This is not about being nomadic in the sense that Gypsies are well-known.

Referring to the Aromanians, when it comes to the phenomenon of transhumance, this means in no way a lack of permanent residence, that place called 'home', merely changing the place to stay, depending on the season, summer or winter, between those two residences.

The Aromanians, even when they were leaving with their flocks of sheep, missing from home for a long time, have always had their settlements where they returned home. Not to mention the fact that traditionally their stone-made homes were above the rest of the surrounding housing, something that all the foreign travelers who had visited the Balkans in the nineteenth century mention.

The Romanic element in the Balkans follows a process of assimilation, most probably due to the location in a Slavic area, but also due to the pressure from the Greeks, especially by many denationalization campaigns. Here we must mention with the greatest sadness that even the Patriarchate of Constantinople had been increasingly becoming the fiercest enemy of the Aromanians.

> *'[...] The Romanians from the South of the Danube, who were swarming with their flocks of sheep all over the [Balkan] Peninsula, starting from Dalmatia up to the Aegean Sea and from the Balkans to the Peloponnese, have been massively denationalized [...] hose who were going to large distances from their native places and settlements, becoming isolated in the middle of the other nations, most of the times finished by denationalizing themselves by forgetting the language. Such processes of losing one's national identity happened mainly in Greece. [...] Starting with the nineteenth century, a main part of them lost their national identity, becoming Greeks by the force of the Church's actions.'*[48]

During A.I. Cuza's time, Anastasie Panu, Chairman of the Council of Ministers, his father being Aromanian, traces the lines of a strategy for the United Principalities, later the Kingdom of Romania, to act coherently in the

---

[48] Capidan, Th., *Macedoromânii. Etnografie, istorie, limbă.* Bucureşti, Fundaţia Regală pentru Literatură şi Artă, 1942, p. 18

Balkan Peninsula. Following his ideas, Romanian schools are created in places with an important Aromanians presence, including the *Romanian High School* in Bitolia.

On the anniversary of 25 years of existence, *Ion D. Arginteanu* speaks about those wonderful people who made this Romanian high school to work away from Romania: *We, the pupils, we thought we were able to touch the sky with our hand and move the earth with our leg.*[49]

> *[...] Romanian High School in Bitola history of over 25 years is the very history of the whole question of Romanian schools in Turkey. [...] In 1878, following relentless insistences from those in question Savfet Grand Vizier Pasha ordered the governors of the provinces, to give the free functioning of Romanian schools. Based on this official permission, Romanian education in Turkey could develop in freedom and take a significant momentum. They suddenly opened several primary schools, as well as in Perlepe, Bitolia, Magarova, Samarina, Nijopole, Furca and others.*[50]

Regarding the way of life of the Aromanians, Capidan makes also reference to a reliable witness, one of the travelers that seems to have better described these things.

---

[49] Arginteanu, Ioan D., *Raportul despre mersul Liceului în curs de 25 de ani,* în Revista *Lumina,* Bitolia (Monastir), oct. 1905, p. 294

[50] Arginteanu, Ioan D., *Raportul despre mersul Liceului în curs de 25 de ani,* Revista *Lumina,* Bitolia (Monastir), oct. 1905, p. 293-294

'[...] *the English traveler William Martin Leake compares Romanian villages in Pindos with the most thriving cities in Greece* [...] *Furthermore, Weigand continues:*
'*Somebody, after seeing the full misery of Bulgarian villages with their clay huts, small and dirty, or the Greek hamlets from Epirus equally miserable, or even mountaineers Albanians' stone homes terribly empty, remains the most enchanted not only by the splendid position and overall impression everywhere they choose, but also by the gorgeous homes and pleasant mood both around and inside them.*'[51]

[51] CAPIDAN, TH., *Macedoromânii. Etnografie, istorie, limbă*. Bucureşti, Fundaţia Regală pentru Literatură şi Artă, 1942, p. 31-32

WHO ARE THE AROMANIANS?

Where is the country of the Aromanians?

For someone who knows something about the Aromanian problem, the answer is simple: nowhere and everywhere!

There is nothing extraordinary here: *anywhere and nowhere.* Let's not forget that the Jews were in the same situation at some point. We gave this example, because there are many similarities between Jews and Aromanians.

Like the Jews, the Aromanians are good at commerce. Like the Jews, the Aromanians give their children the best education. This is the reason why the Aromanians are always at the top of the communities they live in, no matter where they are.

Figure 13 - *Moscopole, Weigand, Die Aromunen Erster Band, p. 67, p. 113,* Digitized by Google

Romanians believe that the Aromanians came from Romanized Dacia, already Latinized, so their country must necessarily be Romania!

Is it Greece? Impossible, if we see their tendency to assimilate the Aromanians. Perhaps and because many of the great Greeks are nothing but Aromanians.

The Vlachs Asan and Petru led the Vlach-Bulgarian Tsardoms, but Bulgaria is not the country of the Aromanians.

If we take into account the fact that the Aromanians have a practically unchanged language for thousands of years and come from ancient Macedonia, then we must necessarily accept that their country is Macedonia. Especially since Macedonia spent 700 years under the Romans, long before Dacia's conquest, it would have been much more likely to be Romanized.

Can this be Albania, for which Romania pleaded so much in 1913 and actively participated in the partition of Macedonia, with so many Aromanians suddenly displaced to different countries? Even Moscopole, the Aromanian flag city, passed to Albania!

In Macedonia, there were always many Macedonian-Romanians, as the Romanians called them. Perhaps God gave them this land.

So what? Romania's policy, which we do not understand today without thinking at the pressures of the Great Powers, was this: the Aromanians should contribute to the formation of a Greater Albania![52]

---

[52] „[...] veţi căuta a apăra, mai înainte de toate, interesele Aromânilor. În acest sens poate fi vorba [...] eventual de o Albanie cât de mare." Ministerul Afacerilor Străine. *Documente diplomatice. Evenimentele din*

What have Aromanians done for the others?

Being anywhere at home, the Aromanians have contributed to the development of the communities they lived in, often being among the distinguished members of these communities, such as the families *Dumba, Sina, Tușita, Sturnari* or *Averof. Emanuil Gojdu* is considered a true reformer of Hungarian criminal law. The loyal and intelligent integration in the society in which he lived went so far that, his policy was not always correctly perceived by the Romanian nationalist radicals, even though he never betrayed his origin.

From the *Dumba* family, Nikolaus was Deputy Governor of the *Viennese Commercial Bank*, a member of the Parliament, an active participant in Vienna's cultural life, a counselor, promoter and patron of many artistic and cultural initiatives.

The *Sina* family also contributed to the creation of institutions of great importance for Hungary. Gheorghe Simeon Sina was with Count Istvan Szechenyi, one of the greatest Hungarian reformers, contributing to the establishment of important institutions of the Hungarian state such as the *Hungarian Insurance Society, the Commercial Academy, the Hungarian National Theater* and even *the Hungarian Academy*. Baron Sina also contributes essentially to financing the construction of the famous *'Bridge with chains'* over the Danube.

Greece had some great Aromanians, such as *Rhigas Feraios*, the precursor of the Greek independence movement, *Georgakis Olympios*, fighter in the revolution

---

Peninsula Balcanică. Acţiunea României. 20 Sept. 1912 - 1 Aug. 1913, Bucureşti, Imprimeria Statului, 1913, p.90

of 1821, *Georgios Stavrou*, co-founder and first governor of the National Bank of Greece, *Ioannis Colettis*, Prime Minister, *Konstantinos Zappas*, who provided the Zappeion Hall and the surrounding gardens, *Georgios Averof*, founder of the Military Academy, *Spiridon Lambru*, historian and politician, *Athinagoras I.*, Patriarch (1948-1972).

Aromanians, people like us

Short and comprehensive, the characteristic feature of Aromanians is to be frugal, sober, active, smart and very skillful, tireless and admirable tenacity in goals and ideas, enlightened, moral and imbued with a sense of duty.[53]

Eminescu, the greatest Romanian poet, as a complex personality, knew whom and how the Aromanians were, and this excerpt proves it very well. It is an article where, when speaking about the Romanians, Eminescu refers to the South of the Danube, i.e. Aromanians:

' Sober, having married instinct and industry, looking at these qualities, Romanians are significantly higher than those who speak Greek; but are inferior to Greek Slavs in spirit and trickery. [...]

---

[53] V. Arion, V. Pârvan, G. Vâlsan, Pericle Papahagi şi G. Bogdan-Duică, *România şi popoarele balcanice*, Tipografia românească, 1913, p.38

> *Merchants and Vlachs guilds are in all European Turkey cities, and even in Hungary the loveAustria, led by love of gain. They are good and rich shopkeepers mostly proves Sina from Vienna, Vlach born in Klinovo.* [54]

Perhaps it would have been better if I would have begun this history of histories about Aromanians with the words of *Henry N. Brailsford* (1873-1958), a famous English journalist of the end of the nineteenth century and early twentieth century.

This journalist was particularly mindful of the Balkans in 1913-1914, being a member of the international commission sent by *the Carnegie Endowment for International Peace* to make a report on the Balkan wars. Brailsford is co-author of the report, which gives a greater authority to his words. Here is how he describes the Aromanians:

> *'They shelter themselves in the Greek Church, adopt Greek culture as a disguise, and serve the Hellenic idea. It is rare to meet a man among them who does not speak Greek more or less fluently and well, but at home the national Latin idiom persists, and their callings, their habits, their ways of thinking make them a nationality apart. They are not a very numerous stock, though without their aid the Greeks would cut a poor figure among the statistics of the Macedonian*

---

[54] Ziarul „Timpul," III, nr. 211, 26 sept. 1878, p. 1

> *races. The so-called 'Greeks' of Monastir*
> *are Vlachs to a man.'*[55]

Stressing that the number of those counted as Greeks in Macedonia would be insignificant without them, Brailsford speaks about Aromanians as *shy* people minding their own business.

> *'If they are shy people they are also singularly*
> *tenacious. A family may be scattered*
> *between Romania and Thessaly, but they*
> *never cease to be Vlachs; and the women*
> *move about among their Bulgarian*
> *neighbors, never abandoning their neat*
> *costumes of navy-blue, more suggestive*
> *of Norway than of the Balkans. They are*
> *the innkeepers and the carriers of*
> *Macedonia.'*[56]

Aromanians and their social values

Generally, to agree exactly what we are talking about, we should mention at the very beginning that the definition of *social values* says that they form a sum of qualities and beliefs common to a specific crop or group of people. These values may be religious, economic, political, etc.

---

[55] N. Brailsford, Henry, *Macedonia: its Races and their Future*, London, 1903, p. 176
[56] Ibid., p. 176

Figure 14 - Aromanian family from Pleasa (right, an Albanian),
Weigand, Die Aromunen Erster Band, p. 113,
Digitized by Google

T. H. S. Escott James Baker says in *La Turquie: le pays, les institutions, les moeurs*, Paris, 1883, that the explanation for the Aromanians resistance to the de-nationalization and graecization process is the firmness with which Aromanian women have defended their identity, being conservative and holding strongly to their mother tongue.[57]

Th. Capidan go further in analyzing the outstanding capabilities of Aromanians, providing a plausible explanation than for their success in business, and in everyday life:

'[...] By working as a means provided by nature and social organization,

---

[57] Baker, James, La Turquie, trad. de J. de Caters, Paris, 1883, p. 147

> *Aromanians, wherever they were [...]*
> *managed to create, in terms of moral*
> *and material, all the social virtues that*
> *they made the elite of the society in*
> *which they lived.'*[58]

Where do the Aromanians' name come from?

Over the years, starting from the name given by the Germans to the *Volcae* tribes, where the ethnonym of *Vlach* comes from, the Aromanians were known under different names, depending on the language in which their names were given.

Aromanians,   Vlachs,   Makedonians,   Vlachs,
Tzintzari, Aromanians, Makedo-Aromanians, Macedo-Romanians, etc. So many names for one people without country. Concerning all the forms under which the Aromanians are known, we may find the explanation Th. Capidan gives as a very good one:

> *'The name of the Macedo-Romanians. After*
> *their definitive separation from the*
> *Romanians in Dacia, which probably*
> *took place between the seventh and the*
> *tenth century., [...] the name 'Rumân',*
> *receiving his prosthesis, a common*
> *phenomenon in their dialect, came to*
> *Arumân, and this, with the fall of 'u'*
> *from the first syllable, Aromân. This is*
> *the only common name for most*

[58] Capidan, Th., *Macedoromânii. Etnografie, istorie, limbă.* Bucureşti, Fundaţia Regală pentru Literatură şi Artă, 1942, p. 26

> *Macedo-Romanians strains. [...] In*
> *addition to this name, the Macedo-*
> *Romanians also have some nicknames.*
> *Serbs and, everywhere, the Southern*
> *Slavs, call them 'Ţinţari', and the Greeks,*
> *'CuţoVlachs'. Albanians in Northern*
> *Albania say 'Gogă'. [...] But the Macedo-*
> *Romanians, having the pastoral*
> *occupation, some of the Balkan peoples*
> *call them 'shepherds' in their language.*
> *Thus, a part of the Turks called them*
> *'Cioban', a name sometimes used by the*
> *Albanians to ridicule.'[59]*

About all those nicknames received from other people speaks also V. Diamandi-Aminceanu:

> *'The Gentiles who, sooner or later, came into*
> *contact with them, gave him special*
> *names and nicknames, such as: Vlachs,*
> *KuţoVlachs, BruzoVlachs,*
> *ArvanitoVlachs, Olachs, Caraguni,*
> *Tzânţari, Boveni, and Boi after the very*
> *testimony of Aravandinos in în*
> *Hronografia tis Ipiru t. II p. 113.'[60]*

---

[59] CAPIDAN, TH., *Macedoromânii. Etnografie, istorie, limbă*. Bucureşti, Fundaţia Regală pentru Literatură şi Artă, 1942, p. 10-11
[60] Diamandi-Aminceanu, Vasile, *Românii din peninsula balcanică*, Bucureşti, 1938, p. 9

Aromanians, religion and Church

According to Western patterns, the personality of the people of the Balkans is a special one. Here, in this part of the world, people are a lot more related to the family than to the community, and this could be immediately witnessed at the Aromanians, perhaps much more than at the others.

Until Martin Luther's Reformation, there were two great religions in Europe: The Catholic and the Orthodox one, somewhat flourishing on those two empires created after the fall of the Roman Empire. Up to a point, Orthodoxy can be confused with Byzantium.

At the beginning of the sixteenth century, the Western Catholic world begins the assimilation of the Protestant values.

In the Catholic world, what counts is the Church, not the individual who practically had access to God only through the Church.

Figure 15 - Aromanian church in Moscopole nowadays
Photo: Eugene Matzota

> '*What does the authority of the Church mean for Catholicism, and how greatly the rights of conscience and individual reason disappear in the face of this authority, said in the most lapidary mode Augustine:... salvation does not necessarily require man's connection with God, but only the keeping of the rules that the Church establishes... [...]* '[61]

Let's see now what Bartholomew of Alverna[62] has to say about the religious practices in the regions inhabited by the Vlachs in the Western Balkans:

> '*[...] they do not believe that the Holy Spirit proceeds from the Father and the Son, like the Greeks, but they do not believe in the authority of the Greek doctors... they do not believe that the holy Catholic and Apostolic Church is one and they say That blessed Peter and his descendants are not the rulers of Christendom, but each apostle and patriarch are with the same authority.*
>
> *[...] they say that every man can save himself in his rite, because it would be impossible for God to have condemned*

---

[61] Ionescu, Nae, *Neliniştea metafizică*, Editura Fundaţiei Culturale Române, 1993, p. 139

[62] *Bartholomew of Alverna* (1379-1382), Vicar of the Bosnian Franciscans

> **all other people who are not Christians.'[63]**

In the Balkans, the relationship between man and God is a warmer and more human one. Moreover, even the ideals and the religious life are different.

The Catholic Church promotes the ideal of the realization of the kingdom of heaven on Earth. Orthodoxy believes that man cannot escape here, in this world, from the effects of the original sin.

Origen, a great theosophist, not just a great Father of the Church, notices that the serpent holds out the apple to Eve, knowing in advance that she will receive it. This, however, proves the existence in man of the bending to sin, even by being human.

> **And then, loving your neighbor as yourself, it would mean no: cherish your neighbor as yourself, but: do not cherish more than others; Or, crueler yet: despise yourself.**
>
> **However, it follows that the true link between men in this life is not the positive bond of love, but the negative and passive mercy of love. The strict logical consequence of Orthodoxy is hermitism.**
>
> **And this is in fact the characteristic of our religious life towards the West: the anchorites, the Sihasters are unknown appearances in the Western Church.**

[63] Dionisius Lasic, O.F.M., *Fr. Bartholomaei de Alverna, Vicarii Bosniae 1367-1407, quaedam scripta hucusque inedita,* „Archivum Francescanum Historicum", LV, 1962, 1-2, p. 66-68

In the East, the ideal is the realization of this Kingdom in us.'[64]

The Aromanians belonged to the world of the Eastern Roman Empire until the fall of Constantinople. After this, they passed under the influence of the Ottoman Empire, but without suffering so much pressure to give up Orthodoxy.

In this world that did not know the Renaissance or the Reformation, closing in immemorial values is probably a proof of respect for traditions, towards the roots. We could say that, when we speak of an Aromanian culture, apart from folk culture, everything that can be called culture is linked in one way or another to the Orthodox Church preserving the same values for so many centuries.

This is how the Aromanian culture was preserved.

---

[64] Ionescu, Nae, *Neliniştea metafizică*, Editura Fundaţiei Culturale Române, 1993, p. 92

DNA, THE FINAL ANSWER?

Does DNA really have all the answers?

Within a study of the genetics of Balkan populations, four DNA-STR systems and 19 classical markers were examined in seven samples: Romanians (two groups), Albanians, Greeks and Aromuns (three groups). The results for the DNA-STR systems have been compared with data from the literature.

The results show four clear separated groups: sub-Saharan black populations, North-African, Japanese and European populations.

The large Balkan populations, except the Greek sample, are genetically more homogenous than the Aromun populations.

A second Neighbor-joining tree based on all 2analyzeded systems, show a particular trend of the Aromun groups, which indicates a particular genetic structure.[65]

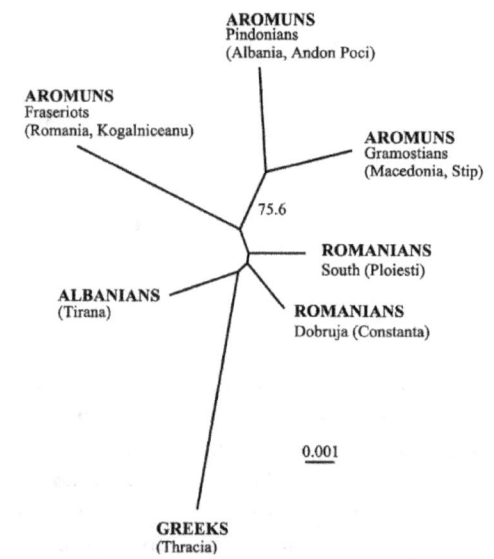

Figure16 - *Neighbor-joining tree of the Balkan populations studied (four DNA-STRs and 19 classical markers)*

---

[65] H. D. Schmidt et al.: South Balkan Populations, Coll. Antropol. 27 (2003) 2: 501–506

The authors of the scientific paper *Paternal and maternal lineages in the Balkans show a homogeneous landscape over linguistic barriers, except for the isolated Aromuns* (2006)[66] treated the whole area of the Balkans as a complex cultural mix, with populations speaking languages that include branches of the Indo-European and Altaic family. In this complex, the Aromanians form, as the authors say, a clearly defined cultural minority.

Seeking evidence of genetic stratification using *'mtADN'* and *'Y'* chromosomes in Albanians, Romanians, Macedonians, Greeks and five Aromanian populations, scientists have tried to genetically test several variants of Aromanian origin. The results indicated *'a possible common origin'* of the Aromanians.

> **'The homogeneity of Balkan populations prevented testing for the origin of the Aromuns, although a significant Roman contribution can be ruled out.'[67]**

Data from Albanians in Tirana, Greeks from Thrace, Macedonians from Skopje and Romanians from Constanţa and Ploieşti, as well as from Aromanians from Andon Poci and Dukasi (Albania), Stip and Cruşova

---

[66] BOSCH, E., CALAFELL, F., GONZÁLEZ-NEIRA, A., FLAIZ, C., MATEU, E., SCHEIL, H.-G., HUCKENBECK, W., EFREMOVSKA, L., MIKEREZI, I., XIROTIRIS, N., GRASA, C., SCHMIDT, H. AND COMAS, D. (2006), *Paternal and maternal lineages in the Balkans show a homogeneous landscape over linguistic barriers, except for the isolated Aromuns.* Annals of Human Genetics, 2006 Jul;70(Pt 4):459-87

[67] *Ibid.*, 70: 469-470

(Republic of Macedonia), Mihail Kogălniceanu Romania) were analyzed. What has emerged is merely demonstrating that genetic diversity in the Balkans is in line with the European model.

'... The main source of genetic differentiation in the Balkans is due to some, but not all, Aromun groups.'

All the Balkan populations analyzed here were genetically homogeneous with the exception of some Aromun samples. [...]

Therefore, it seems that the Aromun populations do not constitute a homogenous group separated from the rest of the Balkan populations, but that they present relative heterogeneity, especially for paternal lineage composition, between themselves. [...]

In spite of their possible historical common origin, the geographical isolation between the Aromun populations analyzed, plus the cultural isolation from their neighbors, may have favored the action of genetic drift on their paternal lineage composition even at the level of binary markers. [...]

However, the repeated pattern in all the paternal lineages found in shared haplotypes between Aromun populations, [...] among the Aromuns, [...] provides further evidence of the

effect of genetic drift in these
populations. [...] [68]

Throughout history, many peoples have had their period of glory: The Egyptians, the Macedonians, the Persians, the Greeks, the Romans, and the Turks, just to remember those geographically close to Europe. Where are the greatness and grandeur of the Ottoman Empire, with a period of glory over five centuries? Could we find that glory in nowadays Turkey, Macedonia, or Egypt?

The confluence of these peoples with others has led to the combination of populations with visible and measurable effects at the DNA level. Today's geneticists are trying to apply new statistical methods to the available data to provide irrefutable arguments for conducting outstanding events over a period of backwardness of about 4000 years.

As they are presented in the latest studies, these events have taken place especially over the past 3000 years. Mutations in the DNA structure are those that provide clear data about a segment's belonging to a particular population or race. The chromosomes inherited by the child from each parent are divided into smaller pieces in successive generations, which allows researchers to find out how many generations have passed from the combination of populations.

[68] *Ibid.,* 70: 469–470

At the beginning of 2014, the prestigious journal *Science* published *A Genetic Atlas of Human Admixture History*, a study by the team led by Simon Myers of Oxford University, Garrett Hellenthal of University College London, and Daniel Falush of the Max Planck Institute for Evolutionary Anthropology in Leipzig. They have detected about 95 distinct populations in the world, including the Romanians, but not the Aromanians.

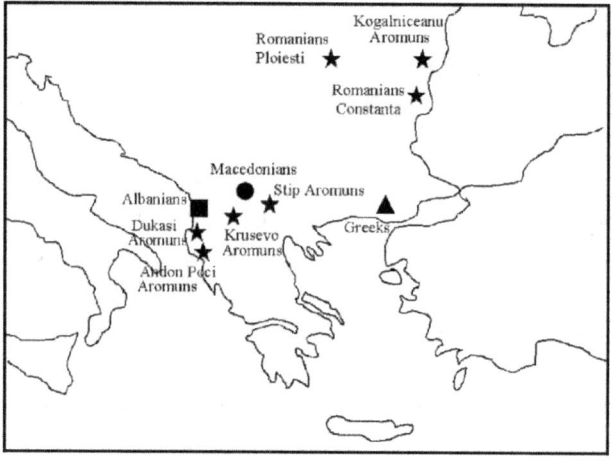

Figure 17 - *Geographic location of the samples analysed. Symbols represent the linguistic classification of the samples: Italic (stars), Slavic (circles), Greek (triangles), Albanian (square).*

Dr. Myers and Dr. Hellenthal said they did not collaborate with historians, and even more than that, as their colleague in Leipzig says, they would not even want to talk to them. Dr. Falush believes that their study is objective, even if the part for the Romanians, for example, relies on only 13 samples. For these scientists, things are simple: just enter some data on one side, and the history goes out on the other side!

Studying their results, we see that the Romanians, for example, mingled in greater proportion with the Finns than the Hungarians, whose language we have known for so long that it is Finno-Ugric.

Now what should we think of all that?

In *Alu insertion polymorphisms in the Balkans and the origins of the Aromuns*, published in 2004 in the reputed **Annals of Human Genetics,** a team led by researcher D. Comas analyzed specific human polymorphisms in the Balkans to elucidate the Aromanian origin, testing three hypotheses:

- *Aromuns are Romanophonic Greeks;*
- *The result of a Romanian Southward migration;*
- *Or local descendants of the Thracians.*

We have analyzed 11 human-specific Alu insertion polymorphisms in the Balkans to elucidate the origins of the Aromuns, a linguistic isolate inhabiting scattered areas in the Balkan Peninsula.

Four Aromun samples (two from the Republic of Macedonia, one from Albania, and one from Romania) and five neighboring populations (Macedonians, Albanians, Romanians, Greeks, and Turks) were analyzed by means of genetic distances, principal components and analyses of the molecular variance (AMOVA).

Three hypotheses were tested: Aromuns are Romanophonic Greeks; the result of a Romanian Southward migration; or local descendants of the Thracians.

> *The analyses show that the Aromuns do not
> constitute a homogeneous group
> separated from the rest of the Balkan
> populations. Grouping by language or
> geography does not explain the genetic
> differences observed in the region,
> suggesting a lack of genetic structure in
> the area.'[69]*

The analyzes carried out collectively by researcher D. Comas show that the Aromanians do not constitute a homogeneous group, separate from the rest of the Balkan populations and do not appear to be especially bound by the Greeks or the Romanians.

> *'Aromuns do not seem to be particularly
> related to Greeks, Romanians, or to other
> Romance speakers. The Aromuns might
> have their origin to the South of the
> Danube river, with extensive gene flow
> with the neighboring populations.*
>
> *The present results suggest a common
> ancestry of all Balkan populations,
> including Aromuns, with a lack of
> correlation between genetic
> differentiation and language or
> ethnicity, stressing that no major
> migration barriers have existed in the*

---

[69] COMAS, D., SCHMID, H., BRAEUER, S., FLAIZ, C., BUSQUETS, A., CALAFELL, F., BERTRANPETIT, J., SCHEIL, H.-G., HUCKENBECK, W., EFREMOVSKA, L., SCHMIDT, H., *Alu insertion polymorphisms in the Balkans and the origins of the Aromuns*, Annals of Human Genetics, Volume 68, Issue 2, , March 2004, p.120

making of the complex Balkan human puzzle.'[70]

DNA tests conclusions

Scenario 1:
Aromuns are Latinized Greeks

'[...] The present data seem to refute this first hypothesis of the Aromuns being Latinized Greeks.'

Scenario 2:
Aromuns and Romanians are descendants of the Dacians

'Even Romanian Aromuns are very differentiated from Romanians, as shown by genetic distances and PC analysis. This could be explained by the fact that after 1920 many Aromuns coming from the South of the Balkans emigrated to Romania, where they settled in the Dobruja region and near Bucharest. This second hypothesis for the same origin of the Aromuns and Romanians, as descendants of the inhabitants North of the Danube and

---

[70] *Ibid.,* p.120

mixed with Romans, can also be rejected.'

Scenario 3:
Aromuns are Descendant of Thracians

'[...] This scenario would be difficult to test since no existent population can be identified with the ancient Thracians. Nevertheless, this hypothesis would imply a close genetic relationship between Aromuns and the rest of the populations living in the Balkans South of the Danube.

The present analysis has shown a close relationship between Balkan populations and Aromuns, although Aromuns of any given geographical region or political country do not present closer genetic distance to their neighbors, [...] .

Nonetheless, a single origin for the Aromuns cannot be confirmed by the present data, given that the genetic heterogeneity found among Aromun samples is similar to that found among a set of linguistically unrelated Balkan populations.

This pattern could be explained if Aromun populations had, in fact, different origins scattered in the Balkans, and had converged linguistically by adopting and preserving a Romance language. Or, alternatively, genetic drift could have erased the traces of a putative common origin.' [71]

---

[71] *Ibid.*, p.126

FACTS AND FIGURES - BEFORE 1800

> *'Not knowing what has happened before you*
> *were born is just being relentlessly babe;*
> *For what is man's age, if our memory of*
> *these facts would not join the previous*
> *generations?'*

Cicero, Orat., chap. 34.

First written evidence of the Aromanian language

Torna, torna, fratre! (τόρνα, τόρνα, φράτρε) is the first evidence of the use of the Aromanian language. These words represent what a soldier is shouting to another one during a night march. This happened in the year 587, during the march of the Byzantine army led by *Comentiolos* against Avars, somewhere near Haemus Mountains, the ancient name of the Balkan Mountains.

Theofilact of Simocatta (570-640), Byzantine historian born in Alexandria, Egypt, recounts this episode in his *'Histories'*. This moment would not have been probably recorded if those words would not have been wrongly interpreted by other soldiers, generating hysteria, repeating those words, over and over. As a result, an unstoppable stampede has begun, with soldiers beginning to flee in opposite directions.

What should be noted here is that Theophylact Simocattes says that the soldier was speaking *'in his own country'* language (επιηοριο τι ψλοττι). The soldier who had used that expression had not done anything else than shouting to another one to turn back because a bag with supplies has fallen down.

Being a text of particular importance, here we quote the passage from the *History of the reign of Emperor Mauriciusy, II, 15*, by Theophilact Simocattes:

> **'Comentiolus marshalled the army, arranged it into a single formation, and permitted it to march. He instructed it to move towards the Astike, to spend the night on guard, and on the morrow to fall on the Chagan like a whirlwind and inflict very great slaughter on his company. But some fate decided to pervert the general's prescriptions; for like a drone it wasted the hives of prudence, and ravaged the general's beelike labors.**
>
> **For when the sun showed its back to grim-faced night, and the beautiful light-bearing torch had shrouded its radiance and given way to nocturnal power, one of the baggage animals shed the load it**

was carrying. It happened that the
animal's owner was marching in front;
those following behind saw that the
beast of burden was dragging in some
disarray its intended load, and ordered
its master to turn to the rear and to
rectify the baggage-beast's miscarriage.
This in fact became the cause of the
disorder and produced a spontaneous
backward rush to the rear.

For the utterance was incorrectly repeated by
the majority, the word was distorted, and
it appeared to indicate flight, as if the
enemy had suddenly appeared before
them and cheated their expectation. The
army fell into tremendous uproar, a
great outcry arose among them, with
piercing shouts everyone cried out to
return, and one man ordered another in
the native parlance to turn to the rear,
amidst utmost confusion, shouting
`Turn, turn', as if a night battle had
unexpectedly come upon them.'[72]

Theophanes Confessor, (752 - 817), a Byzatine chronicler, author of an universal chronicle, tells the same story in his 'Cronographia', a little bit differently, saying that the soldier would have cried *'in parental language'* (τι πατροξορι): *torna, torna, fratre!*

---

[72] *Fontes Historiae Dacoromaniae*, vol. I-IV, vol.II, Bucureşti, 1964-1982, p. 539

> 'A beast of burden had thrown off his load,
> and somebody yelled to his master to
> reset it, saying in the language of their
> parents/of the land: 'torna, torna, fratre'.
> The master of the animal didn't hear the
> shout, but the people heard him, and
> believing that they are attacked by the
> enemy, started running, shouting loudly:
> 'torna, torna'.[73]

The theory that those words were simply a military command cannot be taken into account, because there is no military language using such appellations as 'brother'.

Historian G. Brătianu believes that this sentence:

> 'is an expression from the Romanian
> language, as is has been formed in those
> times in the Balkan and Danube
> regions'; 'They probably belong to one
> and the most important of substrates
> over which language was built'[74].

---

[73] *Ibid.*, vol II, Bucureşti, 1964-1982, p. 605

[74] http://ro.wikipedia.org/wiki/Torna,_torna,_fratre!

Figure 18 - *Mount Athos.*
Photo: Eugene Matzota

The first mention of the ethnonym 'Vlach'

A note found at the Castamonitu monastery on Mount Athos speaks for the first time about Vlachs. The term used here is 'vlahorinhins', i.e. Vlachs from the river Rhynchos valley, which flows into the bay Rendina, in the northeast peninsula Chalcidice.

These 'vlahorinhins', according to this document dating from the ninth century AD, allied with the Slavic tribe of 'sagudats', used to invade Macedonia, attacking Thessaloniki, and reaching Mount Athos.

'In the days of emperor fighters for icons, the nations in the lands near the Danube found an era of anarchy, because the

> *wicked emperors of the Romans waged war against holy icons.*

> *Then 'vlahorinhinis' and 'sagudats' conquered Bulgaria, were spread little by little into different parts, seized Macedonia and finally broke into the Holy Mountain with all the boys and their women because there was no one to stand against and to confront them with war.'*[75]

Vlachs and Macedonians, clearly different

Taking advantage of the anarchy that reigned at the beginning of the tenth and eleventh century in the Byzantine Empire, Vlachs rebelled several times against the Byzantine emperors.

We find then some Vlachs even enrolled in Byzantine armies against which they had fought before, taking part in several expeditions.

Barensis, the chronicler, makes a clear distinction between Vlachs and Macedonians (*Vlachorum, Macedonum* in Lat. In original). Barensis recaps the participants in the expedition undertaken by the Byzantines against Sicily in 1027 during the Emperor Constantine III:

> *'This year has descended Ispo Cheloniti (The Cubicular) in Italy with great army,*

---

[75] *Fontes Historiae Dacoromaniae*, vol. I-IV, vol.IV, Bucureşti, 1964-1982, p. 7

> *namely with Russians, Vandals, Turks,*
> *Bulgarians, Vlachs, Macedonians and*
> *others to conquer Sicily.'[76]*

Kekaumenos about the Vlachs in Thessaly

In the book written between 1075 and 1078, *Cecaumeni Strategicon* (gr: Στρατηγικὸν τοῦ Κεκαυμένου), Byzantine general Kekaumenos reminds about the presence of Vlachs in Thessaly.

'Alexiada' and the Vlachs

Anna Comnena, Byzantine princess (1083 - 1153), one of the first historian women known to us, is the first medieval author we know about writing on the

Figure 19 - *Alexis I Comnenus and Hugues the Great* (Wikipedia Commons)

Vlachs settlements in the Thessaly Mountains.

Anna Comnena says that the Emperor Alexius I Comnenus, before the Lebounion (1091) fight against the Pechenegs, enrolled from the Balkans soldiers gathered from the

---

[76] „Hoc anno descendit Ispo Cheloniti (Cubicularius) in Italian cum exercitu magno id est Russorum, Guandalorum, Turcoarum, Burgarorum, Vlahorum, Macedonum, aliarumque ut caperet Siciliam". Pertz, Georg Heinrich, *Monumenta Germaniae Historica. Anales Barensis*, tom V, p. 53

Bulgarians and those who had a nomadic life.

Here are the texts that speak of Vlachs in *Alexiada*:

BOOK V

When he was close to the territories of Larissa and had passed over the hill of the Cells, he left the public high-road and the hill, Cissabus, so-called locally, on the right and marched down to Ezeba; this is a Vlach village situated close to Androneia.[77]

BOOK VII

He was partly to levy recruits from the Bulgarians and from the nomadic tribes (called Vlachs in popular parlance) and for the rest whatever horse-or foot-soldiers offered themselves from any country.

BOOK X

At night a certain Pudilus, a Vlach nobleman, came in and reported that the Comans were crossing the Danube [...] '[78]

---

[77] Anna Comnena, *The Alexiad*, BOOK V
http://www.fordham.edu/halsall/basis/AnnaComnena-Alexiad00.asp
[78] *Ibid.*

Vlach families on Mount Athos

Between 1100 and 1104, a group of nomadic Vlach shepherds arrived with their families and flocks on Mount Athos. *John Trachaniotes*, a monk, speaks in 1109 about the conflict caused within the holy community he belonged to by the families of Vlachs because women are not allowed on the Holy Mountain to avoid disturbing the monks in the contemplation of the Lord.

The presence of the women of Vlach here led to a dispute in which even the Ecumenical Patriarch Nicholas and Emperor Alexios I Comnenus were engaged. The letters exchanged between the Patriarch and the Emperor remind that there were about 300 families of the Vlachs established near the Holy Mountain, in what is called in the language of the time, 'hamlets.'

The rebellions of the Vlachs

A group of Vlachs and Bulgarians, allied with the Cumanians in the North of the Danube, rebelled in 1185 against Byzantium at *Haemus*, today's Balkan Mountains, and Anchialos, due to the taxes raised by Isaac II Anghelos.

The Allies proclaim a local Wallachian-Bulgarian empire, which becomes the kingdom of the Wallachians and Bulgarians, to reach what we now know as the second Bulgarian Tsardom.

Great Wallachia

Benjamin of Tudela, a great rabbi, starts at 1170 to make a trip to the Orient from Spain, in order to learn more about the Jewish communities there.

He is the author of the first reports of the independent state of *'Great Wallachia'* somewhere in the mountains of the Balkans.

In the manuscript in the *British Museum*, quoted by Bartier in 1735, Ascher in 1840 and Vilalis in 1899, Benjamin de Tudela writes about the people who live there:

'From there it is a day's journey to Sinon Potamo, where there are about fifty Jews, [...] The city is situated at the foot of the hills of Wallachia.

The nation called Wallachians live in those mountains. They are as swift as hinds, and they sweep down from the mountains to despoil and ravage the land of Greece.

No man can go up and do battle against them, and no king can rule over them. They do not hold fast to the faith of the Nazarenes, [p. 18] but give themselves Jewish names.

Some people say that they are Jews, and, in fact, they call the Jews their brothers, and when they meet with them, though they rob them, they refrain from killing

**them as they kill the Greeks. They are
altogether lawless.'**

FROM THAT ERA

- *Vlach Cheddar* is mentioned by
  Byzantine poet *Theodor
  Ptochoprodromos,* who calls it
  Vlachiko tyri in spoken Greek,
  same as in Greece today.

- There are talks about the
  'mantles' produced by the
  Aromanians, the so-called *Cape*
  (Gr. *kappai*).

Despotate of Epirus

In 1205, the Epirus Principality is formed in an
intensely Vlach populated area, headed by *Mihail I
Komnenos Doukas* or *Dukas*, with the name *Comnenus
Ducas* in Latin, which reigns until his death in 1215.

The Epirus Principality, also known as *Despotate
of Epirus*, is one of the states that emerged after the Fourth

Crusade, alongside the Empire of Nicaea and Trebizond, successors of the Byzantine Empire.

In 1337, it was conquered by the Kingdom of Serbia, restored in 1356, to enter the Ottoman Empire since 1479.

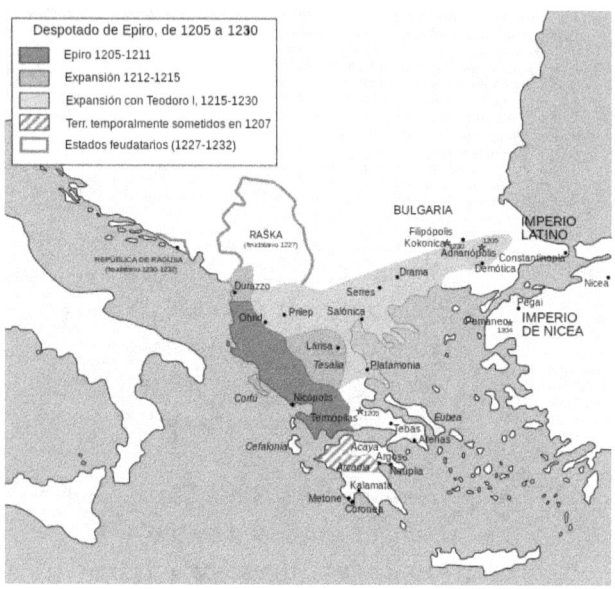

Figure 20 - Epirus between 1205 and 1230[79]
(Wikipedia Commons)

The Chrysostom of Stefan Prvovencani

It is mentioned in the Chrysostom of *Stephen (Stefan) Prvovencani*, dating back to 1220, the submission of the Vlachs from the Serbian kingdom to the

---

[79] "*Epiro 1205-1230*" by Goldorak, based on Winston work. - File:Epirus 1205 1230.svg. Wikimedia Commons - http://commons.wikimedia.org/wiki/ File:Epiro_1205-1230.svg#mediaviewer/File:Epiro_1205-1230.svg

jurisdiction of the archdiocese located at the Zica monastery, near Kraljevo, at the entrance to the Ibar Bay from Central Serbia.

The Vlach-Bulgarian Tsardom

' [...] at 1186, brothers Peter and Asan are appearing on the stage. According to the testimony of all contemporaries without exception, they were Romanians, their people - Romanian, their language - Romanian.

The Romanian element was so widespread in the twelfth to the fifteenth century that the current Thessaly was called the Great Romania, Megâly Blahia; Etolia - Little Romania; Epirus - The Land of the Vlachs; Moesia - White Romania; Current Romanian Country {Actually Ţara Românească, E.M.} - Black Romania (Mauro Vlachia). Finally, the Asanids called themselves in documents, ' Blacorum et Bulgarorum Kings.'[80],

says the greatest Romanian poet and journalist Mihai Eminescu.

[80] Eminescu, Mihai, articol publicat în „Timpul," III, nr. 211, 26 Septembrie 1878, p. 1-2

The Vlachs are decisively involved in the Vlach-Bulgarian Tsardom, where the ruling dynasty was that of the Asăneşti. A. Vasiliev, a great authority in the history of Byzantium, says that the Bulgarians have joined the liberation movement initiated by the ancestors of the Romanians (Blachs). Moreover, he clearly states that Peter and Asan were *Blachs*. About the *Asăneşti Revolt*

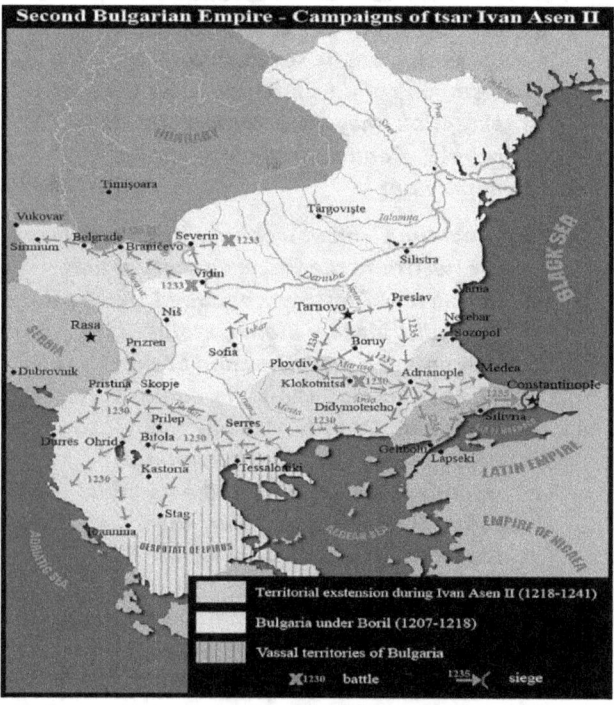

Figure 21 – *Second Vlach-Bulgarian Empire*
Wikipedia Commons

(1185), Niketas Choniates writes when speaking about the marriage of Emperor Isaac Anghelos (1185-1195) with the daughter of the King of Hungary, Bela III (1172-1196):

'But stingy to spend treasury money for the wedding celebrations, he collected them mercilessly from his own lands; and has robbed, from pettiness, other cities in the parts of Anchialos, stealthily, but specially made it to the Roma and to the enemies of the barbarians of Mount Haemus, formerly called Misi, and now called Vlachs.

These, [...] also stood up against the Roma [...]. But especially they made enemies for them and the Romans the barbarians of Mount Haemus, who previously were called Moessians, and now they are called Vlachs. These, [...] also stood up against the Romans [...].

And they were leaders of the evil and agitators of the whole nation, one Peter and one Asan, of the same nation and origin. In order not to revolt without reason, they appear in front of the Emperor, who was in the Kypsela camp, asking to be enrolled in the army together with the Romans and to receive from royal card a domain with some income on Mount Haemus.

But their demand was not fulfilled [...] Beginners of this transgression being those whom we have spoken of, the Emperor opposes them.' [81]

---

[81] Fontes Historiae Dacoromaniae, vol. I-IV, vol. III, Bucureşti, 1964-1982, p. 255

It seems, however, that the Vlachs were not immediately convinced to join Peter and Asan, so there was a need for a kind of divine intervention. By instilling the idea that God, through the holy martyr Dimitrius, would join them in this action, the Vlachs eventually gather around the two future founders of the Vlach-Bulgarian Tsardom.

> *'[...] Thee whole nation takes up arms. [...]*
> *And one of the two brothers, Peter,*
> *crowned his head with a golden coronet,*
> *and made himself and put on the red*
> *shoes.'[82]*

Figure 22 –
*Monument of Ioniță
Caloian in Varna*

The Romanians on the left of the Danube participated too in this rebellion, something explicitly mentioned by Niketas Choniates, who speaks of the passage of the river from North to South by Vlachs (Romanians) and what he calls 'Scythians', namely the Cumans.

Asan and Peter were assassinated by their own boyars in 1196 and 1197 respectively.

Their brother Ioniță Caloian ascended to the throne, strengthened The Vlach-Bulgarian Tsardom, and expanded its borders from the Southern Carpathians to

---

[82] *Ibid.*, p. 255

the Maritza and Rodopi rivers, from the Black Sea to the Vardar River. Even though he was not accepted by the Pope as emperor, Ioniță Caloian manages to obtain his recognition as 'king of the Bulgarians and the Vlachs (interpreted of the Wallachians too)' (*Rex Bulgarorum et Blachorum*), even from the Pope.

Ioniță Caloian has asked Pope Innocent III[83] to be received into the Roman Catholic Churchthe argumentwhe as Emperor of the Vlachs and Bulgarians, invoking as argument his and his people Romanity in favor of this desire.

The resettlement Vlachs of Thrace

Emperor Andronicus II Paleologus decided in 1285 to move the Vlachs from Thrace to Asia Minor, due to the real danger of their alliance with the Tatars. The measure is preceded by excessive taxation of the Vlachs, which leads to the dramatic decrease of their fortunes.

---

[83] „Venerabilului şi Preafericitului Părinte, pontificelui suprem
Eu, Caloian, împăratul bulgarilor şi al vlachilor, îţi trimit voie bună şi sănătate.
[...] Pentru aceea aducem mare mulţumită Atotputernicului care ne-a cercetat pe noi nevrednicii Săi servi după nespusa Sa bunătate şi a privit spre smerenia noastră şi ne-a făcut să ne aducem aminte de sânge le şi de patria noastră din care ne tragem. [...] Iată ce cere împărăţia mea de la Scaunul Apostolic: ca să fim în biserica Romei ca fii adevăraţi ai adevăratei mame. Întâi şi întâi cer de la Biserica Romei, maica noastră, coroană şi cinste ca un fiu iubit, după cum au avut şi vechii noştri împăraţi. [...]"
Tăutu, Aloisie L., *Devotamentul lui Ioniţă Asan către Scaunul Apostolic al Romei*, „Buna Vestire", nr. 1, 1966, p. 7-9

Descriptio Europae Orientalis

In 1308, an anonymous Dominican or Franciscan monk, who has been a missionary for a long time in Serbia, speaks of the Vlachs. His Chronicle, entitled *Descriptio Europae Orientalis*, is accidentally discovered by a Polish scientist, Dr. Olgierd Gorka, in 1913, when he was searching at the National Library of Paris for historical documents about the Crusade.

The original was lost, unfortunately. The author seems to be of French nationality, admirer and counselor of the brother of King Charles IV of France, Carol de Valois - Carolus Valesius, trying to occupy Constantinople.

The anonymous chronicler writes this geography treatise in the spring of 1308, describing not only geographically, but also historically, several countries, including Albania, Serbia, Bulgaria, Bohemia and Hungary, alongside the Byzantine Empire. 'Our Anonymous does not invent anything. He only says the controlled truth. In his words we can put all our trust, in his words we can have all the confidence.

About our countries, not knowing them, he does not speak anything. However, in two passages, he makes some appreciations of the Vlachs in the Balkan Peninsula and their origins, as well as the Roman pastoral and pastoral pastures of Hungary, Romanorum pastores and Romanorum pasca, and

> *also of the Roman principalities in Hungary at the arrival of the Hungarians.*
>
> *Given the objectivity of this anonymous, acknowledged as a true writer, the importance of these assessments will not escape anyone's sight'[84]*

The Anonymous chronicler tells us in the first place that the *Blazes*, as they were known in contemporary French sources, are the same with the *Blachis*. The chronicler also says that they are a very large and widespread people of shepherds, that they were shepherds of the Romans on the territory of present-day Hungary, from where they had to leave on the arrival of the Hungarians.

Besides the fact that the Vlachs are appreciated for their excellent cheese, for the milk and meat they produce, they also have a large, rich country, which was at that time occupied by Carol de Valois's army. The same chronicler also says that the ethnonym *Hungary* is relatively new, the country being called *Pannonia* and *Messia*.

In fact, it should be noted that B. P. Hasdeu had the intuition to say before the discovery of this document that the Aromanians in the Balkan Peninsula are descendants of these *pastores Romanorum* chased by the Hungarians.

---

[84] Trad. Popa-Lisseanu, G., *Izvoarele istoriei românilor, Vol. II, Descrierea Europei Orientale de geograful Anonim*, Tipografia Bucovina, Bucureşti, 1934, p.6

The departure of the Vlachs from the Pannonian Plain could very well explain their presence later in the Northern parts of Slovakia and Poland today.

As we have already said in *the Glossary of Terms*, we are talking about the *Wallachians* (*valaši*) from the Valašsko region, a region located on the North West of the Carpathian Mountains, in the area of Hukvaldy, Valasske, Mezirici, Vsetin, Broumov, Lukov, Vizovice, Zlin.

Fragments of this important information can be found in the original Latin, in the appendices of this book, for those who want to give their own interpretation to those written in *Descriptio Europae Orientalis*. There is also a more consistent quote from the sources of the *History of the Romanians, Vol. II, Description of Eastern Europe by the Anonymous geographer*, which contains fragments from the chronicles of the time, chronicles that support what is said in *Descriptio Europae Orientalis*.

Great Vlachia

Emperor Ioan Cantacuzino gave John Anghelos, through a charter from 1342, the leadership of the province of Great Vlachia, also known as the 'Wallachian Tesalia'.

It is a state of the Vlachs shepherds, which existed in the twelfth and thirteenth centuries, including the Greek area of Thessaly, the central area of the Pindus Mountain, as well as some parts of Macedonia.

From 1204, Great Vlachia was included in the Despotate of Epirus, but quickly succeeded in gaining again independence.

Stefan Dusan's Code

Stefan Dusan's Code makes some interesting references to Vlachs. These are generally acts of royal or feudal donations for Serbian monasteries, texts in which village names are found, as well as names of Vlach families. Here is also remembered the Vlachs' *right for pastureland*.

In other documents, such as the charter of King Stefan Uros to the Hilandar Monastery on Mount Athos in 1302-1309, we can find references to certain regulations of the status of the Vlachs, such as the *laws of the Vlachs*, the so-called *Zakon*.

Vlach special rights

The Turks preserve the special rights of the Vlachs when they reach Europe and become masters over the Balkan Peninsula.

1. High justice will be brought before a judge, according to the Aromanian laws; On this price, the Aromanians will be able to travel freely throughout the empire and will be able to engage in any profession they think fit.

2. The Inner Police will be carried out by the Aromanians themselves, under the responsibility of their own leaders.

3. For the outside police, the Aromanians will agree with the governors of the provinces.

4. No one will interfere in their religious affairs.

5. They will be exempt from taxes; And as a sign of vassality, they will send a gift or a tip to the valide sultan every year.[85]

Ragusa calls the Vlachs

During the war between the Kingdom of Hungary and the Republic of Venice (1378-1381), Ragusa (today's Dubrovnik in Croatia) appeals to the king of Bosnia to send a Vlach contingent to his aid.

Laonic Chalkokondyl on Vlachs

Laonic Chalkokondyl[86], well-known Byzantine humanist, speaks of the identity between the Vlachs South and North of the Danube when he tells the story of Mezid Bey's campaign in the Northern and Western parts of the Ottoman Empire, where at that time the most important name was Iancu of Hunedoara.

Chalkokondyl says that the Vlachs are one nation stretching from the Carpathians to Pindos without being able to tell where their homeland of origin might be.

---

[85] V. Arion, V. Pârvan, G. Vâlsan, Pericle Papahagi şi G. Bogdan-Duică, *România şi popoarele balcanice*, Tipografia românească, 1913, p.32
[86] Laonikos Chalkokondyles, Laonicus Chalcondyles (Gr.: Λαόνικος Χαλκοκονδύλης, (cca. 1423 – 1490).

Vlachs in Hungary, at Miskolc

In 1606, as many Romanian historians say, several Vlach families are settled in Miskolc, Hungary. The Hungarian historians place this moment later, towards the end of the seventeenth century.

Thus, we meet the Vlach from Moscopole, Grabova, Lunca and Civara, who, relying on their native qualities, quickly get rich merchants, managing to obtain the right to exercise long-term retail.

Statuta Valachorum

On October 5, 1630, Emperor Ferdinand II issues *Statuta Valachorum*, a document referring to the Vlachs living between Szavum and Dravum, nowadays Sava and Drava, and regulating relations with the captains of the Vlachs.

The title of this document, in the spirit of time, is very long and begins with something like this:

'*Quemadmodum universa Valachorum Communitas, in trium Capitaneorum supremorum, nimirum Crisiensis, Capronczensis, et Ivanichensis,*' [...] etc.

Vlachs in 'De regno Dalmatiae et Croatiae'

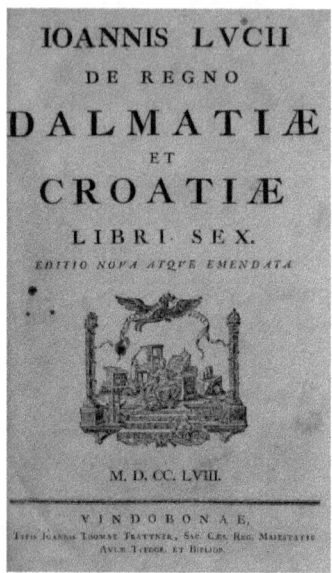

Figure 23 – *De regno Dalmatiae et Croatiae,* cover page
Digitized by Google

The book *DRegnono Dalmatiae et Croatiae*, which includes a chapter *About Vlachs*, appears in Amsterdam in 1666.

The author, *Ioannes Lucius* (Ivan Lučić, 1604-1679), an open-minded historian, without prejudices, wrote the best work of the old Croatian historiography.

Lučić says that '*the name of Vlach, which, to all Slavs, means Romanic and Italic Latin*' was also given to the Romanians in the Balkans', but, because of their enslavement by the conquering Slavs, the ethnonym also acquired the social value of '*dependent man and servant*'.

His texts have allowed different interpretations for the purposes of contesting the value of ethnic term 'Vlach' in the documents.

On the contrary! We see here clearly that Lučić speaks categorically about the Romanic origin of the Vlachs, as well as about the incontestable Roman character of their language.

He also remembers the Vlach origins of Ioniță Caloian, the third king of the Asănești Dynasty, as well as Iancu de Hunedoara:

'Wallachians [Valaches] do not call themselves Vlachs [Vlachos] or Wallachans [Valachos], but Rumanians [Rumenos] and they boast that they descend from the Romans and those who dealt with them testify about it and from their tribe declared Ioniță, the King of Bulgaria and Wallachia, was born, writing to Innocent III;

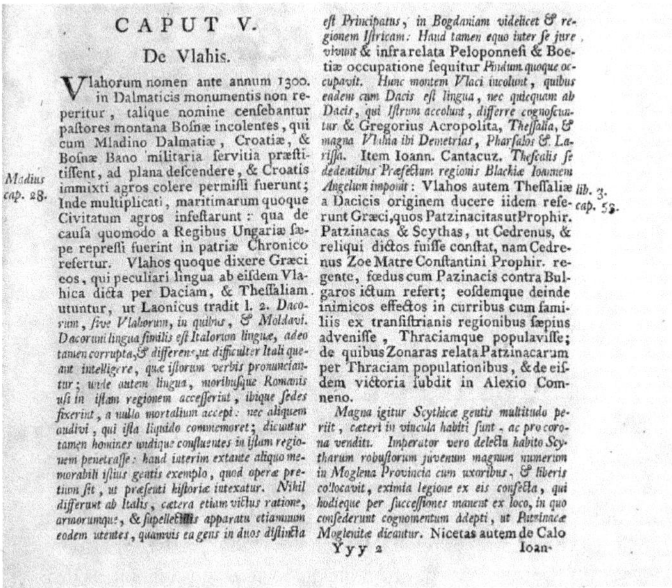

CAPUT V.

De Vlahis.

Vlahorum nomen ante annum 1300. in Dalmaticis monumentis non reperitur, talique nomine cenfebantur paftores montana Boſnæ incolentes, qui cum Mladino Dalmatiæ, Croatiæ, & Boſnæ Bano militaria fervitia præftitiſſent, ad plana defcendere, & Croatis immixti agros colere permiſſi fuerunt; Inde multiplicati, maritimarum quoque Civitatum agros infeſtarunt : qua de cauſa quomodo a Regibus Ungariæ ſæpe repreſſi fuerint in patriæ Chronico refertur. Vlahos quoque dixere Græci eos, qui peculiari lingua ab eifdem Vlahica dicta per Daciam, & Theſſaliam. utuntur, ut Laonicus tradit l. 2. *Dacorum, ſive Vlahorum, in quibus, & Moldavi. Dacorum lingua ſimilis eſt Italorum lingue, adeo tamen corrupta, & differens, ut difficulter Itali queant intelligere, quæ iſtorum verbis pronuncientur; unde autem lingua, moribuſque Romanis uſi in iſtam regionem acceſſerint, ibique ſedes fixerint, a nullo mortalium accepi: nec aliquem audivi, qui iſta liquido commemoret; dicuntur tamen homines undique confluxiſſe in iſtam regionem penetraſſe; haud interim extante aliquo memorabili iſtius gentis exemplo, quod operæ pretium fit, ut præſenti hiſtoriæ intexatur. Nihil differunt ab Italis, cætera etiam victus ratione, armorumque, & ſupellectilis apparatu etiamnum eodem utentes, quamvis ea gens in duos diſtincta*

Mladus cap. 28.

eſt Principatus, in Bogdaniam videlicet & regionem Iſtricam: Haud tamen equo inter ſe jure vivunt & infra relata Peloponneſi & Boetiæ occupatione ſequitur Pindum quoque occupavit. Hunc montem Vlaci incolunt, quibus eadem cum Dacis eſt lingua, nec quicquam ab Dacis, qui Iſtrum accolunt, differre cognoſcuntur & Gregorius Acropolita, Theſſalia, & magna Vlahia ibi Demetrias, Pharſalos & Lariſſa. Item Ioann. Cantacuz. *Theſcalis ſe dedeantibus Præfectam regionis Blachiæ Ioannem Angelum imponit:* Vlahos autem Theſſaliæ *lib. 3. cap. 58.* a Dacicis originem ducere iidem referunt Græci, quos Patzinacitas ut Prophir. Patzinacas & Scythas, ut Cedrenus, & reliqui dictos fuiſſe conſtat, nam Cedrenus Zoe Matre Conſtantini Prophir. regente, fœdus cum Pazinacis contra Bulgaros ictum refert; eoſdemque deinde inimicos effectos in curribus cum familiis ex tranſiſtrianis regionibus ſæpins advenifſe, Thraciamque populaviſſe; de quibus Zonaras relata Patzinacarum per Thraciam populationibus, & de eiſdem victoria ſubdit in Alexio Comneno.

Magna igitur Scythicæ gentis multitudo periit, cæteri in vincula habiti ſunt, ac pro corona venditi. Imperator vero delectu habito Scytharum robuſtorum juvenum magnum numerum in Mozlenæ Provincia cum uxoribus, & liberis collocavit, eximia legione ex eis confecta, qui hodieque per ſucceſſiones manent ex loco, in quo conſederunt cognomentum adepti, ut Patzinacæ Moglenitæ dicantur. Nicetas autem de Calo

Yyy 2   Ioan-

Figure 24 – *De regno Dalmatiae et Croatiae, "About Vlachs" Chapter*
Digitized by Google

And John of Huniade, born between the Wallachians of Transylvania, was boasted to be drawn from the Roman family of Corvina.[87]

Valachi autem hodierni quicunque lingua Valacha loquuntur feipfos non dicunt Vlachos aut Valachos fed *Rumenos*, & a Romanis ortos gloriantur, Romanaque lingua loqui profitentur; quod ficut fermo ipforum comprobat, ita mores quoque eorundem Italis, quam Slavis fimiliores convincunt, ut relati auctores referunt, & qui cum eifdem verfati funt teftantur, ex quorum progenie fe ortum Ioannicius Rex Bulgariæ & Blachiæ ad Innocentium III. fcribens *Rain.* profeffus eft; Ioannes quoque Hunniades inter Valachos Tranfilvaniæ natus *1203. n. 20.* ex Corvina Romana familia ortum ducere gloriabatur.

Figure 25 – *De regno Dalmatiae et Croatiae,* Excerpt from the "About Vlachs" Chapter Digitized by Google

---

[87] "Valachi autem hodierni quicunque lingua Valacha loquuntur seipsos non dicunt Vlachos aut Valachos sed *Rumenos*, & a Romanis ortos gloriantur, Romanaque lingua loqui profitentur; quod ficut fermo ipsorum comprobat, ita mores quoque eorundem Italis, quam Slavis similiores convincunt, ut relati auctores reserunt, & qui cum eisdem versati sunt testantur, ex quorum progenie se ortum Ioannicius Rex Bulgariae & Blachiae ad Innocentium III. scribens professus est; Ioannes quoque Hunniades inter Valachos Transilvaniae natus ex Corvina Romana familia ortum ducere gloriabatur." Lucii, Ioannis, *De Regno Dalmatiae et Croatiae*, Libri sex, Amsterdam, 1666, p.274

Vlachs and Europe

Vlachs, enterprising spirits, are actively involved in communities where they live as merchants and bankers. This involvement in community life appears to be in symbiosis with the Greeks, due to religious affinities.

Even among strangers, the Vlachs retain the consciousness of their own identity. By influencing European trade, the Vlachs contribute to the development of the bourgeoisie throughout the Southeast European and even in Central Europe, that is in Vienna, but also in Venice.

Some Vlachs families from Moscopole, such as the Gojdu family, have arrived in Poland. When Poland is divided between Russia, Prussia and Austria, the Vlachs descend to the South, in Hungary. Here, they become merchants or bankers in the most important cities. The Gojdu family, for example, has settled first at Miskolc, but the Vlachs are also met at Budapest, Timisoara or Sibiu.

Maria Gianni Manarutta and Vlachs in Trieste

Gianni Maria Manarutta, a Carmelite monk with the monastic name *Ireneo della Croce*, author of a book published in Venice in 1698, namely *Historia antica e moderna, sacra e profana della citta di Trieste*, speaking about the residents in the surrounding of Trieste, says they are 'using an own (idiom) and, in particular, similar to the Wallachian blended with various Latin phrases and words...'[88]

Figure 26 – *Dimitrie Cantemir*

Cantemir and Vlachs

Dimitrie   Cantemir writes in his famous book *Descriptio Moldaviae* 1714-1716, about the Aromanian language, as seen by him spoken by the *CuțoVlach*s:

'A language even worse [than those living in Wallachia, compared with the Moldovan speech, E.M.] have the CuțoVlachs living in Rumelia, the border of Macedonia.

They mixed in a surprising way the language of their country with the Greek and Albanian; so that mixed in their speech, Wallachian dialect sometimes snatches of Greek, sometimes of Albanian.

But everywhere, they keep the suffix in Moldavian names [nouns, E.M.] and verbs. In such a spoiled language, they can

---

[88] „(usano un proprio (idioma) e particolare consimile al valacco intracciato con diverse parole e vocaboli latini...)

understand only each other, for no Greek, Albanian and Moldovan is able to understand.

But if all these three would find together in one place and there is a CuțoVlach speaking, then they could certainly understand what the man wants, if each of them would interpret other fragments of his speech.'

Trading company in Miskolc

In 1720, the '*Greek commercial company*' in Miskolc is founded and the Greek and Aromanian were the languages used in this firm. Aromanians of Miskolc were the masters of all wine and wood trade in Hungary and even their exports, mostly to Poland.

The Academy of Moscopole

This new Academy is founded in 1744 in Moscopole, being also called the *New Academy* or *Hellênikon Frôntistêrio.*

The destruction of Moscopole

The city of Moscopole is destroyed between 1760 and 1790 in circumstances that are not clear even nowadays, apparently by Ali Pasha, who had created a kind of local state, against Ottoman power. Besides

Moscopole, other urban centers in the Pindos and Epirus region fall victim to the satrap of Ianina.

Albanian Muslims, jealous of what the Aromanians had achieved, have destroyed the city.

There was some really extraordinary achievement in Moscopole: shopping and banking institutions, manufacturing corporations, public lighting, paved streets and sidewalks, community schools, Academy...

Figure 27 - *Moscopole city crest* [89]

The New Academy was the only higher education institution around, as the Royal Academies of

[89] Wikipedia - http://en.wikipedia.org/wiki/File:Moscopole_coa.jpg # mediaviewer/File:Moscopole_coa.jpg

Wallachia and Moldavia. Moscopole have the only one typography in the Balkans since around 1730.

Pouqueville, Consul General of France in Ianina, says that *'envy and fanaticism have joined together to destroy the work of wisdom.'*

Cousinéry, French Consul in Thessaloniki, agrees, adding that Moscopole was enriched by trade with Germany, making a beautiful city that has become the target of the envy of a pasha in Albania.

He has attacked and plundered the city, therefore, the merchants who have made the fame of this city were pressed to take the path of the West.

Moscopole was a very large city for those days, inhabited exclusively by Aromanians (over 40,000 in 1750, 53,000 in 1788 - according to Pouqueville).

Moscopole had in those times no less than 12,000 houses and 72 churches.

Emigration to the West

After the destruction of Moscopole comes a time when many Aromanian families are migrating to cities in central Europe, an area that was already known to them in terms of trade relations.

Thus, Aromanians get hold of some important economic positions in the Habsburg Empire, being active people, with initiative. Obviously, their fortune was an additional argument.

Miron Costin and the Vlachs

In the second half of the eighteenth century, the Moldovan chronicler Miron Costin already knows about the existence of these Vlachs and about their Romanian origin, as well as the learned prince Dimitrie Cantemir, who, in his book *Hronicul vechimei a romano-moldovlahilor* (1717) recounts several times the Aromanians.

FACTS AND FIGURES - AFTER 1800

The Dictionary of Daniil Moscopoleanul

We know about Daniil Mocopoleanul of Vretos (his true name is *Daniil Mihali Adami Hagi*) that he was 'a priest, a preacher and a teacher. His work, printed at the beginning of this nineteenth century is a four-language dictionary entitled

ÎNVĂȚĂTURĂ INTRODUCĂTOARE
(INTRODUCTORY EDUCATION):

'Do remember that the present vocabulary in four languages was made only and only so that the children of the Mesiodacs would learn the Romans language after they will be especially engaged, with his help, with every word.'[90]

---

[90] Mesiodacs = Vlachs

Aromanian Grammar of Mihail Boiagi

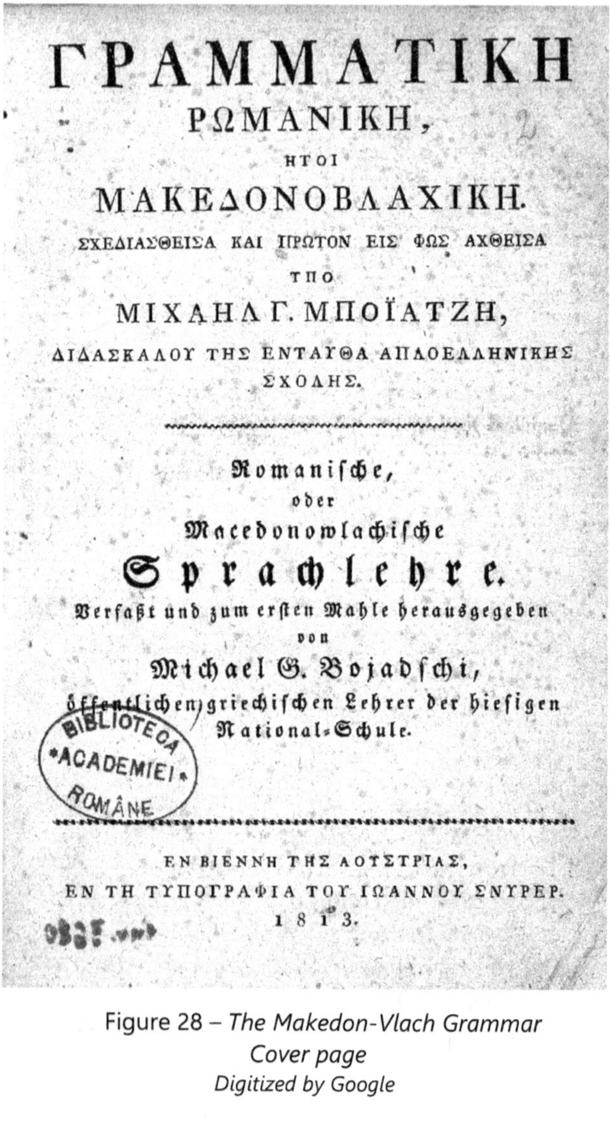

ΓΡΑΜΜΑΤΙΚΗ
ΡΩΜΑΝΙΚΗ,
ΗΤΟΙ
ΜΑΚΕΔΟΝΟΒΛΑΧΙΚΗ.
ΣΧΕΔΙΑΣΘΕΙΣΑ ΚΑΙ ΠΡΩΤΟΝ ΕΙΣ ΦΩΣ ΑΧΘΕΙΣΑ
ΥΠΟ
ΜΙΧΑΗΛ Γ. ΜΠΟΪΑΤΖΗ,
ΔΙΔΑΣΚΑΛΟΥ ΤΗΣ ΕΝΤΑΥΘΑ ΑΠΛΟΕΛΛΗΝΙΚΗΣ
ΣΧΟΛΗΣ.

Romanische,
oder
Macedonowlachische
Sprachlehre.
Verfaßt und zum ersten Mahle herausgegeben
von
Michael G. Bojadschi,
öffentlichen griechischen Lehrer der hiesigen
National-Schule.

ΕΝ ΒΙΕΝΝΗ ΤΗΣ ΑΟΥΣΤΡΙΑΣ,
ΕΝ ΤΗ ΤΥΠΟΓΡΑΦΙΑ ΤΟΥ ΙΩΑΝΝΟΥ ΣΝΥΡΕΡ.
1813.

Figure 28 – *The Makedon-Vlach Grammar*
Cover page
Digitized by Google

Ρωμανογραικογερ- Romanische, griechische
μανικοὶ Διάλογοι. u. deutsche Gespräche.

Α΄. ΔΙΑΛΟΓΟΣ.   Erstes Gespräch.

Μεταξὺ ἑνὸς Ρωμάνε &   Zwischen einem Romanen u.
ἑνὸς Γραικᾶ.   einem Griechen.

Prinzi limba Ro-maneasca ?	Καταλαμβάνετε τὴν ρωμανικὴν(βλαχικὴν) γλῶσσαν.	VerstehenSie die romanische (wlachische) Sprache?
Ne doamne, shi zburescu puçinu, ma ahtantu câtu potu si dau altui si me prindâ.	Ναὶ, αὐθέντα με, ἐμιλῶ ὀλίγον τι, & τόσον, ὥςε νὰ δηλώσω τὰ νοήματά με.	Ja, mein Herr, ich spreche ein wenig, so daß ich mich eben zu verstehen gében kann.
Dicara zburici cu mine.	Ὁμιλήσατε λοιπὸν μὲ ἐμένα.	So reden Sie denn mit mir.
Cu tutâ inimâ.	Μετὰ πάσης χαρᾶς.	Ganz gern.
Cumu treçi zamanea?	Πῶς διατρίβετε τὸν καιρόν σας;	Womit vertreiben Sie sich die Zeit?
Cu preimnarea, câ nu amu verunu lucru.	Περιδιαβάζωντας, ἐπειδὴ ἐδεμίαν δελείαν ἔχω.	Mit Spazierengehen, denn ich habe sonst nichts zu thun.
Nu aveci verunâ cunoashtire in citate?	Δὲν ἔχετε γνωρίμες εἰς τὴν πόλιν.	Haben Sie keine Bekanntschaft in der Stadt?
No, doamne, neci unâ.	Ὄχι, αὐθέντα με, ἀδέναν.	Nein, mein Herr, gar keine.
Eu va si vâ facu cunuscutoru.	Θέλωσας κάμνει γνωρίμας.	Ich will Sie bekannt machen.
Vâ remânu ligatu trâ aistâ.	Σᾶς εὐχαριςῶ κατὰ πολλὰ δί αὐτό.	Ich bin ihnen dafür sehr verbunden.

K

Figure 29 – *The Makedon-Vlach Grammar*
Digitized by Google

The Makedon-Vlach Grammar ('Romanische oder Makedonowlachische Sprachlehre' or 'Γ Ρ Α ΜΜΑ Τ Ι ΚΗ ΩΜΑΝ Ι ΚΗ ΗΤΟ Ι ΜΑΚ Ε ΔΟΝΟ Β Λ Α Χ Ι ΚΗ') of Professor Mihail Boiagi appears in Vienna in 1813, both in Greek and German, with dialect texts, written in Latin.

In German, the word used in the title is 'Romanische', not 'Rumänische', stressing the idea of a people and language with roots in Rome.

The book is structured in a three-column layout, with phrases first in the Makedon-Vlach language, to use the term from the title, and then translated respectively to Greek and German.

FROM THAT ERA

- In 1806, the Orthodox church of Miskolc is sanctified by the bishop of

Figure 30 – *Orthodox Church of Miskolc, The iconostasis*

Buda, Dionysios Popovich. The architect was John Adami. The iconostasis has a height of 16 meters, with icons painted by Viennese Antoniu Kuchelmeister. Greek inscription on the facade says: *'The church of the Holy Trinity was founded in 1785 during the powerful Joseph II, king of Hungary, and was finished in 1806, during the powerful Emperor Francis II, King of Hungary, with the expense of the Vlach brothers from Macedonia'.*[91]

- At Budapest, in 1815, *Atanasie Grabovsky,* Aromanian, the uncle of Andrei Şaguna's mother, encourages all Romanian cultural events; therefore, many books were dedicated to him.

- In 1815, the *Society of Romanian Ladies* of Budapest was founded. The cultural and philanthropic goal is to raise funds for the maintenance of the *Romanian Normal School* in the capital of Hungary, a school that functioned between 1809 and 1888.

The Aromanian women from the aristocracy founded the Society of Romanian Ladies in 1815 in order to make the building

[91] Berényi, Maria - *Viaţa şi activitatea lui Emanuil Gojdu 1802-1870,* Societatea Culturală a Românilor din BudaPest, Giula 2002, pag. 9

> *intended for the school, each woman in the society taking on the task of contributing to the increase of the school fund throughout its lifetime and according to its powers. From this background, the school was to be maintained, teachers paid, and scholars were given to scholarship - which attracted the emperor's praise.*[92]

- The society founded by the great Aromanian colony together with the Transylvanian Romanian residents, a society with educational, cultural-religious purpose, was composed by 80% of the Aromanians. The chairperson is Elena Grabovsky, aunt of Andrei Șaguna. Later comes Marie Roja, the wife of the doctor Constantin Gh. Roja.

The Aromanians of Tudor Vladimirescu

Among those around Tudor Vladimirescu, during the Revolution of 1821, there are also some Aromanians: Farmache and the commander of the Arnauti, serdar Diamandi Giuvara, just to name two of them.

A.D. Xenopol even says in Romanian History. Vol. X, that *'the entire Greek Revolution of 1821 was led and*

---

[92] Hâciu, A., Aromânii. Comerţ. Industrie. Artă. Expansiune. Civilizaţie, Tip. Cartea Putnei, Focşani, 1936, p.558

supported by the Aromanian Armatolians of the Pindus Mountains.'

FROM THAT ERA

- *Dicţionarul în cinci limbi* (The dictionary in five languages) appears, written by the Aromanian Nicolae Ianovici from Budapest, whose subtitle is in the Aromanian: *Diccionariu în cinci limbi: ellinescu, gricescu, romanescu, nemcesu shi madsarescu.*

- Ion Ghica meets at Mrs. De Champy's Salon in Paris, in 1835, with General of Ioan Coletti, Aromanian from Seracu, a doctor of medicine in Pisa and a distinguished person in the struggle for Greece's independence, former Ambassador of the Oriental Cyclades, Ambassador of King Othon in the capital of France.

- Ion Heliade Rădulescu writes in 1838 in his *'Curier de ambe sexe'* magazine: *'The distinction of the Macedonian dialect, we will first examine it in pronouncement, the second in the schematism of the words and the third in the composition of the sentences. All will see that it is the same language and other dialect.'*

- In the *'Spicuitorul moldo-român'* magazine, Morangies saw in 1841 the Aromanians as 'good savages' who *'lived ignorant for centuries in the Pindus Mountains without any ambition but to live a simple and quiet life of thousands and thousands of times happier than those anxious and unsettling peoples looking for happiness everywhere, without finding it anywhere.'*

- *'Organulu Luminarei'*, a magazine from Blaj edited by Timotei Cipariu, published in 1847 four pieces of Aromanian prose.

- *Scrisoare din țara țânțarilor (Letter from the mosquitos' country)* by C. Dăscălescu appears in Iași in 1847 in the Latin-Cyrillic Transitional style.

- Nicolae Balcescu writes in 1848 from Paris, where he was in exile, to Ion Ghica, confessing to him, among others:

'I had the determination, coming to Constantinople, to sit down between Wallachians in the Balkans, because we reckon that these Vlachs will fit sometime where they are.

If you could send a good man there to report on their moral and political status, then

we would look for a school and thus give
work to many starving young people.'

Memories and manifests

- *Dimitrie Bolintineanu* addressed in 1853 a memorandum of the Sublime Porte, namely the Grand Vizier Fuad Pasha in the Aromanians question.

- In the same year, *I. Bratianu* addressed a memorandum to Emperor Napoleon III of France, as did Anastasie Panu in 1863, requesting the autonomous organization of the Aromanians and demanding the cultivation of their nationality through Romanian schools and churches.

- *Dimitrie Cozacovici* launches manifests to Romanians and Aromanians in Romanian, Aromanian and Greek, inviting them to support the founding of Romanian schools for Aromanians.

- Dimitrie Bolintineanu, Christian Tell, C.A. Rosetti and C. Bolliac send in 1863 to Albania, Epirus, Thessaly, and Macedonia a manifesto to Romanians in Macedonia, in Romanian and Greek, urging them to revive their national consciousness.

- *Costache Negri,* a poet and, since 1859, the representative of the United Principalities at Constantinople, a man who is constantly interested in the fate of the Aromanians, advances in 1860 his first official address to the Sublime Porte concerning the improvement of the situation of the Aromanian minority.

First Macedo-Romanian Committee

In Bucharest, the first Macedo-Romanian committee was formed in 1860. The leader of this committee is the Aromanian Dimitrie Cozacovici. Other members of the committee are Goga brothers from Clisura, Mihail Niculescu from Târnovo, Zisu Sideri and Toma Tricopol from Cruşova.

The Committee addresses in the same year a memorandum, elaborated by Panu, to Napoleon III, which states that the number of Aromanians would be 4 million.

The first Romanian school in Turkey

In 1864, the first Romanian school was founded in Turkey at Târnovo, near Bitolia, by Dimitrie Atanasescu, who transformed his own home in school. The number of pupils varies between 20 and 80.

However, in November 1864, began the persecutions lead by the Greek clergy: Benedict of Pelagonia (1854-1869) denounces Atanasescu to the authorities and confiscates 1,000 alphabets. After only 18 hours of detention, he is released, provided he obeys Benedict.

Atanasescu tries to found a school in Bitolia in 1865, but although the Patriarch has pledged *'respect for all the nationalities that profess the Orthodox Faith of the East'*, he is expelled from the Manastir vilayet.

After three years, he can finally return to Târnovo, since Constantinople had issued in the meantime a provision to the provincial authorities that Romanian teachers, because Atanasescu was no longer the only one, had to be left alone.

He remains at his post for 35 years, publishes numerous schoolbooks, in a total circulation of 21,000 copies (distributed free of charge to children) and intervenes for the opening of Romanian schools in other villages.

Northwest of Bitolia, in Gopes, the Aromanian D.G. Cosmescu has founded the second Romanian school in Turkey.

FROM THAT ERA

- An institute where young Aromanians are formed as teachers and founders of schools opens in Bucharest between 1865-1870. They are selected with the help of Archimandrite Averchie. Young learners in Bucharest were obliged to return home after graduation. Ioan

Caragiani also deals with the same problem.

- In September 1868, Jacob and Vanciu Dimonie founded a Romanian school in St. George's district in Ohrid, in the building of the Greek communal school. The teacher is Gheorghe Tomaras, who, immediately cornered by the Greeks, had to ask for the support of Vice-Consul Oculi.

- The communal school passes into the definitive use of the Romanian party in 1871, and Tomas's successor, Filip Apostolescu, a former student of the St. Apostles' Institute, then works in this school for 17 years.

- Mihai Eminescu says in an article published in *Curierul de la Iaşi* on December 6, 1876: *'There is no state in Eastern Europe, there is no country from the Adriatic to the Black Sea that does not contain pieces of our nationality. Starting with the shepherds in Istria, from the morlacs in Bosnia and Herzegovina, we find step by step the fragments of this great ethnicity in the Albanian Mountains, Macedonia and Thessaly, Pindos as well as in the Balkans, Serbia, Bulgaria, Greece and up Beyond the Dniester, close to Odessa and Kiev '.*

- In December 1879, the Romanian representative at Constantinople expresses his intention to take action in Vienna so that Austrian-Hungarian consular officers take over the protection of the Aromanians in all areas where there is no Romanian consul in office.

- Intervening alongside the Sublime Gate, Romania obtained the right to set up Romanian schools for Aromanians in 1879.

The Great Vizier against the Greek Patriarchate

The great Vizier, Savfet Pasha, gave an order on September 12, 1878, concerning the school and religious situation, which was addressed to the valleys of Thessaloniki and Ianina:

'The Sublime Porte was informed, on the one hand, that Romanians in Epirus, Thessaly, Macedonia want to learn a book in their own language and establish schools, and, on the other hand, the Greek clergy, emboldened by the spirits of the unclean, they may have different obstacles and even persecute Romanian teachers.

Considering that in our Empire no one is allowed to hinder the pursuit and full exertion of cult and school teaching, you

**will be willing to make known to civil
servants that you are under the lordship
of your Excellency that they do not have
to press on any of the inhabitants and do
not forcibly object to any exercise of
worship and teaching, and when they
ask for necessity, even to defend the
Romanian teachers.'**

The Macedo-Romanian Culture Society

In 1879, the *Romanian–Macedonian Culture Society* is founded, recognized as a legal person by the law passed by the *Corpurile Legiuitoare* (Law Entities), promulgated by the High Decree of the Government no. 1298, April 15, 1880.

A board of 35 important personalities of the time heads the Society. Among the members, we mention here Titu Maiorescu, Vasile Alecsandri, CA Rosetti, Dimitrie and Ion Ghica, D. Bratianu, I. Campineanu, Gh. Chitu, Chr. Tell, M. Ghermani, Dr. Kalinderu, DA Sturdza and I. Caragiani.

It is decided to focus on the following five objectives:

1. Recognition of the Company as a legal entity.
2. Establishment of an Aromanian bishopric in connection with the autocephalous Church of Romania.
3. Establishment of a boarding school for scholars from Turkey.
4. Obtaining a grant from the Ministry of the Interior for the publications of the Society.

5. Support from the Church for propaganda and attracting new members.

The Romanian High School in Bitolia

As part of the strategy set by Anastasie Panu, Romania is making the necessary steps to open a Romanian high school in the Balkans, which is happening in 1880. Ion D. Arginteanu says on the anniversary of a quarter century since the establishment that the history of the high school also the history of the whole issue of the Romanian schools in the Ottoman Empire.

Let us remember that, since 1878, the Grand Vizier Savfet Pasha had instructed the provincial governors to allow the functioning of the Romanian schools:

> **'Civil servants must give the Romanians full freedom to practice cult and school education, and in case of need to defend and even help Romanian teachers.'**

Several schools for primary education are opened in Bitolia, Perlepe, Magarova, Samarina, Nijopole, Furca etc, immediately, in 1880, when the school inspector was Apostol Mărgărit.

> **In 1880 [...] the first Romanian gymnasium in Bitolia was established, under the direction of V. Glodariu, in whose chest**

the whole fire of patriotism in
Transylvania was boiling..

The world was different then, it was not as
fanatical as it is today. The first
students, Take Pârța, Constantin
Athanasiu, Sideri Carbonari were pupils
from the Greek school. Two years later,
the school gained considerable prestige,
while Father I. Georgiade - Murnu was
the director, Andrei Bagav and
Constantin Cairetti, as well as V.
Petrescu.

About the first, we students, I thought they
were able to touch the sky with their
hands and move the earth with their
feet. The figures of these three apostles
remained a holy trinity in the minds of
the students who had the happiness to
hear their magic word.[93]

During the next two years, the Romanian high school in Bitolia becomes a respected school, having as director Father I. Georgiade - Murnu.

At the 25th anniversary, the number of graduates is 218, of which 19 became professors, 53 institutes, 26 traders, 56 students, 11 doctors, 14 officials, 5 lawyers, 2 reviewers, 3 pharmacists, 3 pedagogues, 1 officer, 1 photographer, 3 farmers. The photographer was one of the famous Manachean brothers.

Considering the total number of pupils enrolled in the first grade of high school, 534, adding pupils from other schools and enrolled in different classes, it results

---

[93] Arginteanu, Ioan D., *Raportul despre mersul Liceului în curs de 25 de ani*, Revista Lumina, Bitolia (Monastir), oct. 1905, p. 293-294

that in the Romanian High School in Bitolia nearly 609 pupils have passed in 25 years.

The Greek Patriarchate fights the Aromanians

The Greek Patriarchate does not recognize the right of the Romanians to pray in their own language and addresses in 1882 a formal order to the archbishops of Macedonia and Epirus to do everything to punish the Romanian priests and close the Romanian schools.

From that moment on, the divine service was officiated only in Romanian in the Aromanian churches, not in Romanian and Greek.

Romania obtained in 1889 from the Ecumenical Patriarchate the right to use the Romanian language in the Romanian churches. The Sublime Gate confirms the right to use the Romanian language in the Romanian churches.

FROM THAT ERA

- The Sultan accepts the entry of an Aromanian in the Reform Commission through a law dated September 13, 1903, following the *Muerzsteg* agreement.

- Alexandru Lahovari, the Romanian envoy to Constantinople, once again presented to the Patriarch a series of Aromanian wishes, insisting first of all for the installation of a Wallachian bishop instead of the Greco-Orthodox Church of the Ohrida-Cruşova diocese.

- In 1904, a Romanian consulate was opened in Ianina, where A. Pădeanu, former consul in Bitolia, was installed.

- The attacks of Greek gangs begin with the killing of the Aromanian D. Liuka of Pisoderion in November 1904. Other leading Aromanians, priests, teachers are beaten or threatened with death.

- In 1905, a public library in Aromanian was established at Bitolia, following the efforts of N. Papahagi and N. Batzaria.

The situation is getting worse

The Greek gang attacks continue. In February 1905, in the village of Phlampouro, five Aromanians and Albanians are cut into pieces with the ax. The situation is becoming more and more serious, as Ghica, the Romanian envoy to Athens, is recalled in the country, being too much on the Greek side.

The arrival of two Romanian school inspectors at Ianina and their arrest under the pretext of provoking a struggle between the nationalist and the Greekophil Aromanians led to the start of the diplomatic machine in April 1905.

The Irade dedicated to Aromanians

A Romanian ultimatum was submitted calling for the two school inspectors to be released and compensated by May 23, 1905. In the same ultimatum, the recognition of the Aromanians *'as a distinct nationality with equal rights to those of other non-Muslim nationalities'* is required by an official Irade.

Just one day before the deadline, on May 22, 1905, Sultan Abdul Hamid has issued the famous Irade, recognizing for the Aromanians, as Romanians, the quality of *Millet*, that is, of *distinct nationality*, as well as the right to self-administer in their own communes. The text, published in *'Le Courrier des Balcans'* no. 39 of 1905, is this:

> *'His Imperial Majesty the Sultan, [...] considering the demands made at the feet of the Imperial Throne by his Romanian subjects, he pleased to order that, by virtue of civil rights, they would enjoy the same rights and title as the others, and their community to designate mayors according to the current legislation and the Romanian members to be equally admitted in the administrative councils and the facilities to be granted by the imperial courts to the teachers employed by certain committees for inspecting and carrying out the formalities dictated by the*

Empire laws for the opening of new school establishments. '[94]

The bizarre position of Romania

The Irade does not influence the action of the Greek gangs, unfortunately. Only in 1905, 23 people were killed, households were destroyed, together with sheep flocks, schools, books and books of worship were burned, without counting anything that has been stolen.

For these reasons, in the context of national indignation in Romania, the state of conflict between Romania and Greece would have led to a war if the two countries had a common border, as Lahovari, the war minister of Romania, said then. In November 1905, due to the Greek massacres, the possibility of arming the Aromanians was considered. King Carol I is the one who expressly prohibits any kind of terrorist action.

Nicolae Iorga, the great historian and politician, said in 1928 that it would have been better for Romania to support the Aromanians by helping to preserve their identity.

Now, at the beginning of the century the way of thinking was different...

---

[94] „Le Courrier des Balkans", no. 39, 1905

BALKAN WARS

The outbreak of conflict

The last major confrontations in Europe were the two great World Wars. With just a little before the first one, there were the two Balkan wars, as a kind of rehearsal. Each of the countries involved, amid the emergence in the nineteenth century of the feeling of belonging to a nation state, wanted a piece of the dying Ottoman Empire.

'*All countries entered World War I each had their interests without thinking of consequences, which have resulted in 10 million deaths. General Lyautey, Resident General of France in Morocco declares, 'They are completely crazy! A war between Europeans means civil war, the monumental idiocy that the world has ever done'.[95]*

World War I seems to have been triggered by the assassination in Sarajevo in June 28 1914 Franz Ferdinand's Austrian archduke and heir to the Austro-Hungarian throne by Gavrilo Princip, a Bosnian Serb nationalist student. The assassination is not only an escalation of tensions already created in Europe and especially in the Balkans, because the Peace of Bucharest of 1913 brought not peace, but only a sort of truce.

The increasing influence of the German Empire, the fall of France after the defeat in 1871, the complex system of alliances, treaties, agreements, secret or not, the proclamation of the German Federal Empire (the Second Reich) under the 'Iron Chancellor' Otto von Bismarck, all these were demanding blood.

The geopolitical situation after the Russo-Turkish War of 1877-78 gave rise to the expansion desires, to a Greater Serbia or Greater Bulgaria etc., as was the case from country to country.

Now, the Serbs already had obtained some territories after the war. The Greeks received Thessaly in 1881. On the other hand, Bulgaria obtained Eastern Rumelia in 1885. All these countries, together with Montenegro, they all wanted parts of Ottoman Rumelia.

---

[95] "Larousse", "Istoria lumii de la origini până în anul 2000", ed. Olimp, București, 2000, p. 499

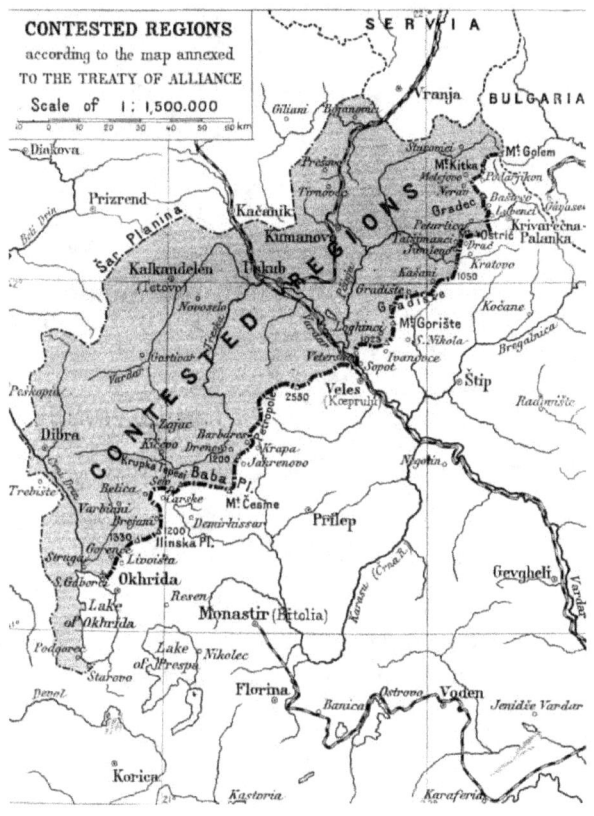

Figure 31 - *Regions disputed by Balkan countries.*[96]

---

[96] *Report of the International Commission To Inquire into the Causes and Conduct of the Balkan Wars,* Carnegie Endowment for International Peace, 1914, p.45.

Figure 32 - Balkan countries' aspirations.

'It was hardly any part of the territory of Turkey in Europe that is not claimed by at least two challengers.'[97]

---

[97] *Ibid.*, p.38.

In large, the territories concerned were Macedonia, Albania and most of Thrace, taken from the Turks in the early conflicts by the member countries of the *Balkan League*, namely Bulgaria, Montenegro, Greece and Serbia. Later on, they couldn't reach an understanding when it came to dividing them.

> *Unity of interests of Russia with Austria in the defense of their common enemy, Romania, will make at least this thing to agree. No victorious Austria, together with Romania, against Russia will not give - it is believed, later - Romania's Bessarabia; nor victorious Russia, together with Romania, against Austria, would give Transylvania to Romania (but only in exchange for Moldova whole).[98]*

As seen from the analysis made on 15 December 1912 by Vasile Pârvan, when nobody knew how things will develop, the Great Powers are very important in the Eastern European complex situation, not only in the Balkans. Today, when we read the telegram sent by the Ministry of Foreign Affairs only two weeks later, we can see that Pârvan was right (the telegram is cited few pages on, for comparison):

---

[98] V. Arion, V. Pârvan, G. Vâlsan, Pericle Papahagi şi G. Bogdan-Duică, *România şi popoarele balcanice*, Tipografia românească, 1913, p.8

> *Romanian nationalists, having probably these things in mind, they thought an alliance between Romania and the Balkan powers, by means of which can take at least from Austria the Romanian trans Carpathians lands.*
>
> *To secure this friendship, so-called nationalists said they enjoyed sincerely all the progress of Bulgaria and Serbia, having not the slightest pretension to rectify the border, or having any political interference in the Balkan Peninsula, contenting themselves with a simple episcopate and ensuring freedom of the Romanian language in school and church for Romanians in Macedonia, as all trends and fully Romanian ideals have to be supported and developed North and not the South.* [99]

As our great historian Nicolae Iorga said, megalomania has been always induced easily by the Great Powers to Balkan peoples in search of their identity. In the collective memory of the peoples, there always has to be a Greater Albania, a Greater Serbia, a Greater Bulgaria, and so on...

Incidentally, today those who want a Greater Albania have territorial claims of 40,000 km2 with a population of about 5,000,000 inhabitants, for example. Similarly, Greater Serbia would be today just kind of new Yugoslavia's Tito, in which the missing part would be only Slovenia and Croatia.

---

[99] *Ibid.*

For a better understanding of the interests of the Great Powers in the Balkans and, in particular, how they tried to distort the actual situation on the ground according to their interests, let us pay attention to what the Romanian great historian Nicolae Iorga has to say.

Describing in 1916, during a conference held at the Romanian Athenaeum shortly after the Balkan wars, the manner of handling used to change the balance of forces in the Balkans, Iorga says:

'Certain Powers sent researchers who were not only beloved by ethnography, geography and history, but were beloved by certain policies, namely their country particular purposes. Two very distinct categories of political science researchers or political scientists came, some of which coming from Vienna, others from Petersburg. Did Austria need an Albania? It always finds researchers who reduced the lifetime of the West Balkan Peninsula to Albanese senses. We know these things from latest times also. (Hilarity.)

They were striving to seek places where Albanians have never been or are only in very recent times, following migration committed at the end of the eighteenth century and early nineteenth century. On the other hand, Russia was sending one Venelin, a Slav native of the Habsburg Monarchy, but which had been established in Russia, together with the Russian army, and he walked in all parties seeking as many Bulgarians as to

*oppose the Serbs whom Austria believed
to be able to use.*

*Besides, the liberal powers, France and
England had every incentive to give as
many senses in the Balkans to Greeks[100]*

Aromanian Question

The two Balkan wars were very important for the Aromanians fate, who were scattered in all of the Balkan states.

Aromanians policy choices were relatively diverse, as an authority in the field, namely Peyfuss, says in *Die Aromunische Frage*, based on reports of Austrian consuls.

Apparently, Aromanians with Romanian sentiments supported the establishment of an Albanian state, while those having pro-Greek feelings opposed.

Romanian media, as often in times of war, brought into the limelight the issues of the brothers in other Balkan countries.

We are not talking necessarily about the Macedonia area, an area so coveted by its neighbors, but about the brothers in Serbia. In general, the media drew attention to all areas where the brothers were subjected to a process of loss of identity.

It was about, first of all, the well-known anti-Romanian attitude of the Greeks, but also by the attitude

---

[100] Iorga, Nicolae, *CE ÎNSEAMNĂ POPOARE BALCANICE.* Conferință ținută la Ateneul Romîn în ziua de 13 Decembre 1915, Neamul Românesc, Vălenii-de-Munte, 1916, p. 20

of the Serbs versus the Romanian elements in the Crajna area, under a long process of assiduously serbization.

It is interesting to study the maps of the 'Report of the International Commission of Investigation of the Causes and Conduct of the Balkan Wars,' in 1914, where one can see both the views of Serbs on Macedonia and the surrounding regions, and that of the Bulgarians. Just a quick preview can see how history was written and seen differently from Sofia or Belgrade.

For example, on the Balkans map seen by Serbs, we may see Romanian populated areas (Cutzo-Vlachs), while Bulgarians transform them into Turks, together with the Russians, as being representative for certain areas.

Moreover, one cannot understand how could the Bulgarians be neighbors with Mount Athos, far away from any community of Bulgarians...

Romania's demands for Aromanians

The Kingdom of Romania had been in a position of neutrality, since the beginning of the first Balkan War, forced also by the position of the Great Powers, who could not stand a Christian country attacking another Christian one, namely Bulgaria, when she fights one Muslim one, i.e. Turkey.

Romania to Bulgaria requirements were primarily related to ensuring the rights of Romanians, i.e. of Aromanians from the territories occupied by the Bulgarians.

Secondly, the situation imposed after the Treaty of Berlin in 1878, explicitly the loss of Silistra, with its fortresses, had to be corrected, together with an amendment on the border to the Black Sea, around Turtucaia, up above Varna.

Here is how the Romanian ambassador to Sofia, D. J. Ghika, expresses a position that has the Kingdom of Romania in this issue, according to the available options:

'*Sofia, 4/17 November 1912.*

On the fate of the Rumanian peoples of Macedonia, even those that the main lines of division between the allies place under other dominations than the Bulgarian domination, I believe that we can draw upon the Bulgarian Government both from His sympathies for an autonomous Albania and the deep crack that has already occurred between Bulgaria and Greece [...]

From what has been entrusted to me... Monsieur Guéchoff is uneasy about the Greek claims, which he finds in petto exorbitantes, especially compared to their meager military successes. The Greeks intend to have Janina and all Epirus, including strictly Albanian regions. As, moreover, it seems that the Greeks should be attributed to the country around Bitolia (Monastir, in original, E.M.), *we could demand of Bulgaria that she should second our views* [...] *D.J.*

GHIKA[101]

In the context of frequent diplomatic consultations between Romania and other Balkan countries, especially Bulgaria, as with the Great Powers, the position the ambassadors of Romania had to defend was not very complex. The only serious problem could have been related to the reparations for losses incurred in the mouth of the Danube after the Treaty of Berlin in 1878.

We have to recall that Romania had to return to Russia the counties in Southern Bessarabia, some historic territories belonging to Romania: *Cahul, Ismail* and *Bolgrad*, which are nowadays part of Ukraine. Despite the fact that had a crucial participation in the campaign of freeing Bulgaria, although received Northern Dobrogea, the Danube Delta and the small Island of the Snakes, Romania had been forced to cede historical territories regained only a few years before, after the Treaty of Paris 1856.

Therefore, the guidelines of the Ministry of Foreign Affairs, later even by Maiorescu, as prime minister and foreign minister, often relate to the South of the Danube brothers' issues. Likewise, the fate of the monks of Mount Athos is not forgotten.

What is noteworthy, also, is Romania's position towards creating a stronger Albania, which contain important territories for Aromanians history.

---

[101] Ministerul Afacerilor Străine. *Documente diplomatice. Evenimentele din Peninsula Balcanică. Acţiunea României. 20 Sept. 1912—1 Aug. 1913*, Bucureşti, Imprimeria Statului, 1913, p.9

It is striking, also, that Moscopole, the famous city of the Aromanians, specifically its ruins, is nowadays in Albania. Not so far away neither from Macedonia, nor from Greece, but in Albania.

'Minister for Foreign Affairs to the Minister Plenipotentiary in London. Bucharest, Saturday, 28 December 1912.

In case of the possible participation of Romania at the meeting the Ambassadors, you will seek to protect, above all, the interests of the Aromanians. In this sense, it can be an autonomous Macedonia and Albania, a possibly bigger Albania. Your dealings will in persuading the Balkans, and especially Greece, to respect Aromanians schools and churches and must not impede in nothing the establishment of a bishop of their own.

And the interests of Romanian monks from Mount Athos could form the subject of your observations, guided by the attached special note, for this particular thing, in more intimate contact with the Russian Ambassador.[102]

Much clearer is the telegram sent by Maiorescu to the Romanian Plenipotentiary Minister in London, with detailed names of places, as well as with requirements relating to the language used in

---

[102] *Ibid.*, p.90

administration in those areas. More than this, one may find here even what should be written for this purpose in the new Constitution of Albania.

'The telegram to the Minister of Foreign Affairs Minister Plenipotentiary in London. Bucharest the 22 March 1913.

Would you so kind to ask that the Macedo-Romanian towns [...] to be incorporated in Albania. Insists on the principle that all the localities where the Romanian form the majority, the Romanian will be the administrative language, the same as all Romanian churches and schools; this principle should be in the Constitution of Albania. T. Maïoresco'[103]

Following same ideas of the configuration of the new borders, we have the following document sent to the famous Foreign Office, before signing the Treaty of London, 30 May 1913, which marks the end of the first Balkan conflict.

Memorandum submitted to the Foreign Office, following the request of Sir Edward Grey.

---

[103] „Veuillez demander que les nombreuses communes macédo-roumaines du Pinde entre Samarina et Metzovo soient incorporées à l'Albanie. Insistez sur le principe que dans toutes les localités où les Roumains sont en majorité, la langue administrative soit roumaine, de même que dans toutes les églises et écoles roumaines; le principe général devrait être inscrit dans la Constitution de l'Albanie." - Ibid., p.90

London, 27 March 1913.

The Romanian Government sees with satisfaction the creation of an independent Albanian state, especially that he hoped that, by the measures guaranteeing Great Powers will want to take, the national individuality of many Romanians that will be incorporated, will be saved. For this purpose, the limits of future Albania should be developed so that not only the state is immune to any further problems with its neighbors, but also the Romanian population, which is the most compact in the Southeast Albania to be kept intact within the limits of the Albanian state.

A population inhabits the region between the towns of Janina, Metzovo, Grebeni and Mount Gram, with a Romanian majority that can be estimated at more than 80,000 inhabitants and is grouped into 36 villages and towns, the most important being: Samarina, Avdela, Perivoli, Crane, Labanitza, Seracu, Perivoli, Laïsta, Leshnitza, Breaza, Turia, Medjidi, etc.

It should be noted that the two slopes of Pindos, from Mount Gramos to Agrafa are occupied mostly by Romanians.

Part of this population was annexed by Greece after the Treaty of Berlin. Romanians protested then against this

> *annexation. It would be unfair to allow the new compact separation of Romanian torso and attach to other countries than Albania. Romania considers that their national individuality is best preserved in an independent Albanian state under the guarantee and control of the Great Powers, whose limits should be set in such a way as indisputable as possible to avoid disturbances in the future in these countries [...]*[104]

If in the First War, Bulgaria, Montenegro, Greece and Serbia fought together against the Ottoman Empire, which finally lost the war, during the second one, Montenegro, Greece and Serbia, which have joined the Ottoman Empire and the Kingdom of Romania, defeated Bulgaria.

The First Balkan War ended with the signing of the Treaty of London, on 30 May 1913. Peace was short, with Bulgarian troops attacking on June 16 Serbian and Greek positions.

The War was motivated by the situation in Macedonia, where Bulgaria wanted to annex an area as large as possible. Macedonia, as other areas where Aromanians are significantly present today, such as Eastern Rumelia, Albania and Thrace, were conflict zones. The Ottoman Empire lost almost everything he had in Europe for 500 years.

Fights for influence over Macedonia have negative repercussions for Aromanians, with insufficient

---

[104] Original French document available at ANNEX.

support from Romania. We have to remember that, while Bulgaria and Greece supported armed gangs undermining the authority of the Ottoman Empire in the area, King Charles I opposed expressly actions of the same kind from Romania.

In an area where the Great Powers had special interests, Aromanians were left virtually to manage on their own, without vital support, which was seen at the Peace of Bucharest, which ended the Second Balkan War.

On the other hand, there is a special sort of acting in the Balkans, where rules are meant to be broken. For example, although Serbia had promised Bulgaria most part of Macedonia, the situation after the war proved that Serbia and Greece had other plans.

Anecdotally, to understand more clearly what we mean by this particular *Balkan-style*, I will give the example of the exchange regarding the cities of Thessaloniki and Serres.

In the first war, the Greek troops entered Thessaloniki just one day before the Bulgarians.

Now, Bulgarians have asked to be allowed to have their entry in Thessaloniki with a battalion. Greeks have accepted, if they were allowed to enter a unit in Serres.

Wonderful… The only problem was the real size of that small battalion of Bulgarians, which proved to be, in fact, a flood of 48,000 people!

The Peace of Bucharest, 10 August 1913

The *Peace of Bucharest*, when Romania incorporated Quadrilater, is a sad moment of remembrance for Aromanians. A Romanian MP spoke about the importance of Silistra, saying, '*even if we lose all Aromanians, Silistra offsets*'! Sure, it was just an opinion, perhaps, but anyway...

The chance to have a place of their own, which could very well be historic Macedonia, with so many Aromanian communities, was lost in the madness of the South Dobrogea land. The same land that became after 1920 their home...

Unfortunately, the territories where Aromanians were at home for hundreds of years, if not thousands of years, were desired by Serbia, Bulgaria, and especially by Greece. We must also add to these countries the new Albania, with the active contribution of the Romanian state.

The status of the Aromanians after the end of the two Balkan Wars are left up to governments of Balkan countries. Therefore, the Aromanian delegates, Murnu, Vala and Papahagi, went in vain to London to talk to the representatives of the Great Powers, as in vain were welcomed by all those representatives.

All that was achieved, as N. Misu, Romania's representative, says, is just some assurance that the Kingdom of Romania will be the one that will best expose and support more persuasive their problems in front of the Great Powers.

> **Minister Plenipotentiary in London by the Minister for Foreign Affairs. London, 14/27 March 1913.**
>
> **President of the Council,**

I have the honor to inform you that Messrs Murnu, Vala and Papahagi delegates Macedo-Romanians arrived in London and that, following the actions of mine, they were received by Sir Edward Grey and all Ambassadors [...] Everywhere there was the best reception for our nationals. They were listening carefully to them and were expressed their sympathy for the noble cause that came to plead for.

All ambassadors also took the view that it is better to entrust their cause to the Romanian Government, which is heard in the European Council and is better able to defend their interests, by the prestige Romania enjoys and by means of which the Royal Government It has to deal with the fate of its nationals in the Balkan Peninsula.

In the intervening what I did for receiving the said men, I always noticed that their mission has no official character and that their words cannot commit at all the Romanian Government.

> *Macedo-Romanian delegation left today*
> *London for Berlin, satisfied with what*
> *they had heard.*[105]

Everybody wins? Well, not exactly...

Greece increases from 2,660,000-4,363,000 inhabitants through new territories annexed: Epirus, South Macedonia, Thessaloniki, Kavala, among others, areas where Aromanians were always present. Secondly, South Macedonia is given to the Greeks, missing the chance to restore Macedonia as a state similar to ancient Macedonia, not necessarily the same. Greece extends to the Northwest, including Ianina.

Bulgaria, who had started the war and finally lost it, gets also a piece of Macedonia, reaching a population increase of 129,490, Apud Carnegie Report, p. 418. Even if Bulgaria failed to take Macedonia, there is a magnification of the territory, even with the loss of Quadrilater (6,960 km² and 286,000 inhabitants).

Serbia also takes a piece of Macedonia, namely Central Macedonia with Ohrid, Bitolia, Kossovo, among others, thus increasing the population of more than 1,500,000 inhabitants, almost doubling the country surface (!!!), from 48,300 in 87,780 km² according to the

---

[105] Ministerul Afacerilor Străine. *Documente diplomatice. Evenimentele din Peninsula Balcanică. Acţiunea României. 20 Sept. 1912—1 Aug. 1913*, Bucureşti, Imprimeria Statului, 1913, p.93

Handbook for the Diplomatic History of Europe, Asia, and Africa 1870-1914[106].

The fundamental flaws of the Treaty of Bucharest were that:

1) Limits that were fired had little connection with the nationality of the inhabitants of the affected regions, and

2) Punishment to Bulgaria, probably deserved given the great challenge of starting the Second Balkan War, was so severe that she could not accept the treaty as a permanent arrangement. While Serbia, Greece and Romania cannot get rid of much of the blame for the character of the Treaty, should not be forgotten that their action at Bucharest was largely due to the settlement imposed to Balkan states by the great powers at the conferences in London. [107]

To Aromanians, this division of Macedonia, with Romania being also a part of it, as a signatory to the Peace of Bucharest, as a witness to the process of scattering their South of the Danube brothers in several countries, is a real tragedy. Macedonia remains until today the apple of discord in that area, especially for Greece, which was unhappy with what received.

---

[106] Anderson, F. M., Hershey, A. S., *Handbook for the Diplomatic History of Europe, Asia and, Africa 1870-1914*, Washington, Government Printing Office, 1918, p.440
[107] *Ibid.* p.440

Once more, the Great Powers used their influence in the spirit of maintaining over time a state of belligerency, according to the old principle *Divide et impera.*

The agreements initialed only by an exchange of letters between Maiorescu, who was then the Romanian PM, and heads of other Balkan governments have not been respected, apart from Greece.

As a result, despite so many commitments of respect for Romanian schools and churches existing in their territories, everything ended up not only by the abolition of schools, but also by the prohibition of using Romanian language in churches.

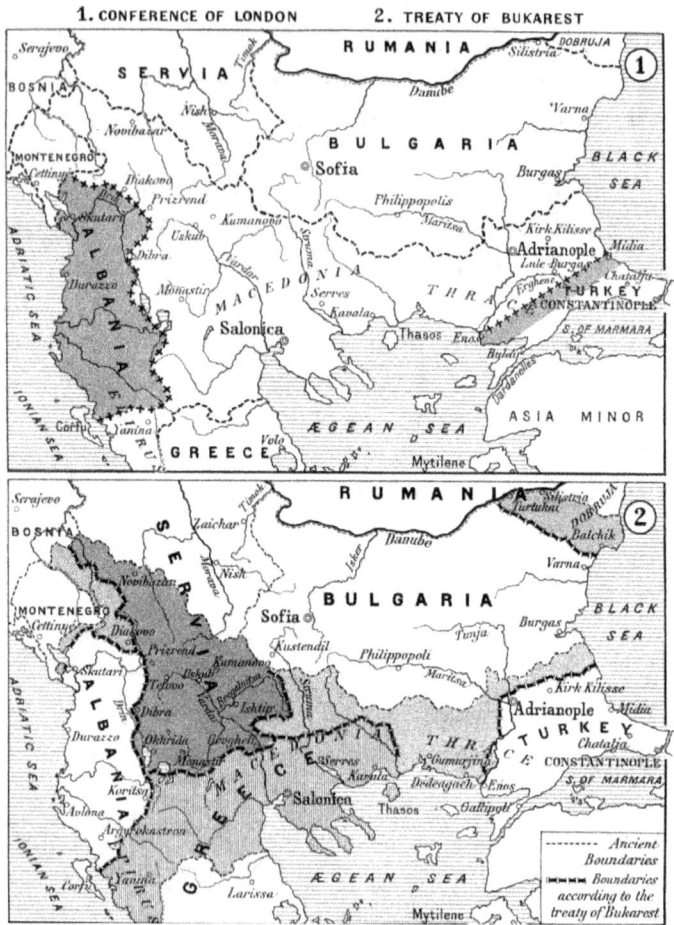

Figure 33 - *Changes in the Balkans following the two conferences, one in London, and one in Bucharest*[108]

[108] *Report of the International Commission To Inquire into the Causes and Conduct of the Balkan Wars*, Carnegie Endowment for International Peace, 1914, p.38.

196

EUGENE MATZOTA

AFTER THE TREATY...

Principality of Pindos

During July-August 1917, there are some efforts to create the Principality of Pindos. This attempt of the Aromanians in Pindos comes amid a favorable context created by the presence of the Italian troops in the area.

The temporary arrival, as we shall see later, of the Italian troops in Pindos is received with the greatest enthusiasm by the Aromanians. They truly believed that this area had been given definitively to their Italian brothers.

It is a time of great ferment, national states are being created and empires break down, the nationality principle being the principle that makes those empires that seemed to be eternal to become nothing more than a memory.

In the summer of 1917, in the heart of Pindos, at Samarina, was proclaimed the existence of a Romanian country, as written in the telegrams sent by the representatives of Romanian settlements

both to Ion I.C. Bratiano and foreign ministers of France, Britain, Italy, Russia, USA.' (Apud St. Brezeanu, Gh. Zbuchea (coord.), Românii de la sud de Dunăre. Documente).

There is nothing to blame when people lost among the Greeks are trying to create in that particular context this Principality of Pindos, with the capital of Aminciu (Metsovo), a form of state autonomy under the protectorate of Italy.

It is convened a national congress on 27 July at Samarina, where Aromanians decide to create this new state formation. In a new meeting, it is decided on July 30, 1917 the election of the tricolor flag decorated with the Roman wolf.

The rapid departure of the Italian troops, immediately followed by the coming of Greek military units, interrupts the Aromanians' aspirations. Romania has a clearly diplomatic position, the Romanian consul in Ianina recommending through a circular to the Romanian school teachers in settlements in the Pindos Mountains *'caution and reserve.'*

On July 23, the director of the Romanian School from Turia, N. Nibi uses exalted expressions such as *'Better dead, than Greek yoke'*.

On August 30 / August 31, the Consul of Romania and the Italian one in Ianina, acting within diplomatic customs, dissociates himself from the action of Aromanians in Pind, saying to Alcibiades Diamandi that *'their step was wrong, inappropriate, since not approved by anyone and cannot be supported in any part'*.

FROM THAT ERA

In 1918, documents mention in the fourteenth and fifteenth centuries, near Dubrovnik area, a number of 67 Vlach hamlets. Out of 67 hamlets, 15 are located in the mountains of eastern Dubrovnik, specifically in the mountainous region of Durmitor, Grahovo, Niksic, Trebinje, Bileca, to Neretva. The rest were hamlets in Montenegro. The Vlachs' numbers were so large that the entire region located between Ragusa, Trebinje and Niksic is named in a Council act of the small town of Ragusa *'ad Vlacos,' 'partes Vlachiae'* in the late fourteenth century.

Peace with Bulgaria

The peace treaty with Bulgaria is signed on November 27, 1919, at Neuilly. Keeping with the disastrous policy towards the Aromanians, the Romanian government does not ask, unfortunately, to include clauses on an exchange of populations, which would have given a better fate for the Aromanians.

In the general context of the Balkan policy, so to speak, the states created after the dissolution of the Ottoman Empire in Europe start a real offensive against the Romanian element by closing schools and churches protected by the Treaty of Bucharest.

The situation gets out of control, leading to the passage of special laws against the Aromanians, as in Bulgaria, which has disowned them through the law for forests and grazing.

Almost immediately follows a massive uptake of Aromanians in Bulgaria, Yugoslavia and Albania. Apparently, Greece was the only one that tolerated Romanian schools and churches. This was not due to any openings towards the Aromanians' communities, but to the mediation by Romania of its relations with Turkey.

Greco-Turkish war

The Greco-Turkish War between 1919 and 1922 causes a mass exodus of Greeks from Asia Minor to Greece, but also to Bulgaria. This change of the economic framework strongly affects the Aromanians' economic interests in the two countries, leading to a mass emigration to Greater Romania.

Little Entente (1920 – 1921)

The Little Entente is a defensive policy organization between Czechoslovakia, Romania and the United Kingdom of Yugoslavia, created in order to maintain the territorial integrity of the three States before Hungary's revisionist claims trends and the actual restoration of the Habsburgs.

The Romanian Government, through Tache Ionescu, was vigilant in order to guarantee the rights of the Aromanians in Yugoslavia, by signing the Treaty of the Little Entente on June 7, 1921.

FROM THAT ERA

- I.I.C. Brătianu, Prime Minister of Romania, decides in 1925 that the Aromanians economically persecuted by Greeks, Bulgarians and Yugoslavs should be displaced to Caliacra and Durostor counties.

- The shepherding crisis of 1925 occurred, caused by the movement of the Greek refugees from Asia Minor to Macedonia after the Greco-Turkish war of 1919 and 1922, and also by the population exchange between Greece and Turkey. To this crisis is added the land reform made for allotment of land to those coming from Asia Minor, which leads to degradation and destruction of the great land area, so necessary for grazing. This is the bad context in which occurs massive emigrations of Aromanians in Greater Romania, between the years 1925 and 1935.

- Nicolae Iorga, head of the Government of Romania between 1931 and 1932, as a reliable defender of Aromanians, tries to improve the situation of settlers from Southern Dobrogea. At a conference held at the South-East European Institute, Iorga says:

> *'We, the people here (in Northern Danube)*
> *together with them (from South of the*
> *Danube), represent all Eastern Romanity;*
> *we have a duty not to forget them [...].'*

• In June 1933, during the Congress of the Cultural League, Nicolae Iorga holds a speech for granting Romanian citizenship to settlers who, just like any Romanian citizen, should enjoy the product of their work and take part actively in the state affairs, talking with much respect about the Aromanian colonists in Dobrogea:

•

> *'You have brought to Romania qualities that*
> *anyone must recognize and cherish. In*
> *the land of your ancestors, you have not*
> *been enslaved to anyone... You have*
> *been masters in your homes... Masters*
> *you must feel here too'!*

• On October 1933, Aromanian settlers speak to Al. Vaida Voevod, calling for a concerted action by the Romanian authorities to protect life and property in the Durostor and Caliacra counties against Bulgarian nationalist attacks organized by the *Revolutionary Committee of Varna.*

- On September 1935, citizenship certificates to over 5,000 householders are granted in ceremonies at Silistra and Bazargic, making them citizens of Greater Romania.

- A large number of colonized Aromanians are forced to leave their households in 1940 on disposal counties in the Quadrilater, and take refuge in Romania.

The Communists and 'the Aromanian Question'

Now, enter the communists...

The communist state resolves simply the 'Aromanian Question' in 1948 by giving up schools and churches which functioned in the Balkans, selling off all their assets, following a decision taken by the horrific woman called *Ana Pauker*[109].

---

[109] Ana Pauker (born Hannah Rabinsohn; 1893 – 1960) Romanian communist leader, foreign minister. The world's first female foreign minister in 1947. *Time* magazine called her "The Most Powerful Woman in the World".

GREAT AROMANIANS - BEFORE 1800

We have gathered here some of the most important names of the less known Aromanians who have made history in the communities they lived in and where they have always been among the distinguished members.

History has often made them Greeks, perhaps because they spoke this international language of that time.

They felt everywhere at home, accepting the situation, in the spirit of specific tolerance.

We are now doing an act of normality by quoting Thede Kahl, who is undoubtedly an authority in the field, making a remarkable gesture when he sums up names of great Greeks (Aromanians):

The role of Aromanians in Greek history, represented by numerous artisans, national heroes, politicians, intellectuals and clergy, is of great importance for their identity as part of the modern Greek people.

Especially their number among Greek benefactors is often accentuated by politicians and historians.

Indeed, the list of examples of Aromanians in Greek history is impressive:

Aromanians are among the fighters of independence, Rhigas Feraios (1757-1798, precursor of the Greek independence movement),

Georgakis Olympios (1772-1821, member of 'Filiki Etaireia', fought in the Revolution of 1821),

Theodoros and Alexis Grivas (1797-1862 and 1799-1855, respectively, the leaders of the armed forces of revolutionaries; they were some of the best known Mecena,

Baron Georgios Sinas (1783-1856, heirloom and superior officer of the Austro-Hungarian Empire, founder of the Academy of Athens Simon Sinas (1810-1876, banker, railway magnate, baron of the Austro-Hungarian Empire, first Greek ambassador to Vienna),

Nikolaos Stournaris (1806-1853, founder of the Metsovio Polytechnic School in Athens),

Georgios Stavrou (1785-1869, cofounder and first governor of the National Bank of Greece);

And were also among the most important politicians Ioannis Kolettis (1773-1847, minister and later prime minister 1844-1847),

Stergios Doumbas (1794-1870, Vienna banker),

Konstantinos Zappas (1814-1892, equipped the Zappeion Hall and the surrounding gardens),

Georgios Averof (1818-1899, founder of the Military Academy, restorer of the Panathenian Stadium),

Nikolaos Doumbas 1830-1900 (banker),

Spiridon Lambros (1851-1919, historian and politician),

Athinagoras I. (1886-1972, Patriarch 1948-1972),

> *Evangelos Averof Tositsas (1910-1990,*
> *Minister of Foreign Affairs 1958-1963,*
> *Minister of National Defense 1974-*
> *1981).[110]*

---

[110] Kahl, Thede, *Aromanians in Greece: Minority or Vlach-speaking Greeks?*, http://www.farsarotul.org/nl27_1.htm

Nicoliță, head of the Vlachs

The Byzantine Emperor Vasile II, the Bulgarocton, named in 980 *Nicolița*, actually called probably *Nicoliță* or *Niculiță*, the chief of the Vlachs from Hellada's *Thema*[111]:

'For this theme of Hellas, besides the
provincial militia, we are told of two
special bodies of great importance, each
to be placed under the command of a
servant, the excavators of the theme, and
the Vlachs. The latter were to be special
to this theme of Hellas, where the Vlach
population was so numerous.

Those of these barbarians who 'raised
military service to the Empire probably
formed a special corps under the
command of a chief designated by the

---

[111] *Thema* is the name given to a province of the Eastern Roman Empire, around 600 A.D., when the official acts began to be written in Greek

> *basiliscus, sometimes named for life as*
> *here Nikolitzes.'[112]*

At the time of Emperor Constantine, the Eleventh Duca in 1067, there is a Vlach uprising under the leadership of the Nicola grandson of that one called *the Duke of Hellada Theme*. The Nicoliță family had become so respected that, when Tsar Samuil occupied Larissa, that it was the only one spared.

St. Nicodim of Tismana

Archimandrite, the founder of the Vodița, Vișina, and Tismana monasteries, considered the protector of Oltenia, was born in Prilep or Kosovo, in a family of pious Aromanians.

His mother, who was one of the ruler of Basarab

I daughters, makes this future holy man a relative to Serbian, Transalpine and Bulgarian royal families.

When he comes Knyaz, his relative Ștefan Lazăr, gives Tismana Monastery ten Romanian villages in Serbia. Nicodim receives his monk name at the Hilandar Monastery on Mount Athos, where he even gets to the forefront of the community.

Figure 34 - *St. Nicodim of Tismana*

In 1369, Nicodim crossed the Danube at Orșova by swimming, or, as

---

[112] G. Schlumberger, *L'Epopée Byzantine*, Hachette, Paris, 1896, tom I. p. 635

the story of one of his miracles goes, floating on the monk race. Then, in 1371, he founded the Vodița monastery ('Apița' or 'Apoasa'), following a vision.

In 1377, the Tismana Monastery was built. Towards the end of his life, he retires to Prislop Monastery in Transylvania, probably because of a conflict with Mircea the Elder. Here, he uses calligraphy to write a *Tetravanghel*, the oldest book dating from Wallachia.

Chalkeus, the great figure of the Enlightenment

Ioan Chalkeus (1667 - between 1730 and 1740), a man of culture, a philosopher, one of the great figures of the Enlightenment, born in Moscopole.

After his studies in his native town, he continued his journey to Rome, where he converted to Catholicism, becoming director of the Greek school in Venice (1694-1703 and 1712-1716).

Returning to Moscopole as a teacher, he has Teodor Cavallioti among his students, who is also one of the most important names in the Greek Pantheon.

First book in the Aromanian language

Teodor Anastasie Cavallioti (1728-1786), priest, pedagogue, philosopher and Aromanian linguist.

His work, *Protopirie (First Teaching)*, Venice, 1770, a reading book, is the first written testimony of the Aromanian language. This reading book for elementary classes is written in Modern Greek and consists of a collection of biblical texts.

At the end of this reading book there is a vocabulary of 1170 Greek words that are translated into Aromanian and Albanian.

In fact, the vocabulary was also published by Johann Thunmann in the *Untersuchungen über die Geschichte der östlichen europäischen Völker*, Leipzig, 1774.

Constantin Ucuta and 'New Pedagogy'

Priest and Aromanian pedagogue born in Moscopole, eighteenth century, former dean priest in Posen, at that time in Prussia, Poznan today in Poland. He published his *New Pedagogy* or *Light Abecedary to teach young children Romanian writing* (Vienna, 1797), a book written using the language spoken in Moscopole.

Ucuta militates for the learning and cultivation of the Aromanian dialect.

FROM THAT ERA

- The Aromanians in the Gramos Mountains come to Megala, today's Livadia.

- Clisura and Moscopole become important centers, controlling trade with Venice, Brindisi and Vienna. The legendary wealth of Moscopole has brought a great cultural momentum.

Metropolitan Filaret

Metropolitan Filaret (1735 - 1794) becomes a monk at Căldăruşani monastery, where he is abbot for a while.

As an archimandrite, he is part of the delegation sent in 1770 to Petersburg to Catherine II of Russia.

He was elected Bishop of Râmnic in 1780, then, in 1792, he became the Metropolitan of Ungrovlahia. As a bishop of Râmnic, he published more than 25 books at his expense, among which Ienachita Văcărescu's *Grammar* in 1787.

St. Joseph the New of Partos

Jacob, on his name of baptism, was born in 1568 in the town of Ragusa in Dalmatia, today's Dubrovnik, from Aromanian parents who were engaged in trade at sea.

After his father's death, Jacob moves with his mother in Ohrid, studying at the Monastery of Our Lady.

He then goes to Mount Athos at Pantokrator Monastery, where he is named as the monk Joseph. In 1650, the Patriarchate of Constantinople sent him to become a metropolitan of Timisoara.

Figure 35 - *St. Joseph the New of Partoș*

Three years later, he retires to Partos Monastery near Timisoara. His akathist is saying that *'he was the adornment of the hierarchs, the exorcist of all passions, the salvation of the faithful, the beautiful praise of Timisoara and the honor of the Partos.'*

Velestinlis, the father of the modern Ellada

Rigas Velestinlis (1757 – 1798), his real name being *Riga Fereu*, from the Aromanian village of Velestin, is the ideological father of the modern Ellada.

Also known as *Rigas Feraios* or *Rhegas Pheraeos*, he is a poet, thinker and revolutionary, born in a wealthy family of Veleştin.

Figure 36 - *Rigas Velestinlis*

GREAT AROMANIANS - AFTER 1800

Coletti, the ideologist of the Great Byzantium

Great statesman, prime minister of King Othon I, also known as Ioan Coletti the healer (Doctor) (1773-1847), is the ideologue of the 'great idea', namely the restoration of the Great Byzantium as modern Greek nation state.

He was born in Seracu, near Ianina, which was the capital of the *'Confederation of*

Figure 37 - *Ioan Coletti*

Aromanian Communes' in the seventeenh century, consisting of 42 localities, some 40,000 souls. She works at Ianina as a healer, private doctor and secretary of Ali Pasha of Ianina. In 1822, he took part in the 'Greek National Assembly' in Epidaurus, where the Greece's independence is proclaimed, being selected minister. Between 1835 and 1843 he was appointed ambassador to Paris.

The Grammar of Mihail G. Boiagi

Mihail G. Boiagi (1780-1842), a philologist, is known for his work (*Romanic or Makedo-Vlach Grammar*), printed in Aromanian, Greek, and German (*Grammatiki romaniki itoi macedonovlaki - Romanische oder macedonowlachische Sprachlehre*, Vienna, 1813).

> *'My work, which is the grammar of the Macedonian-Romanian language, as the Romanians in the South of the Danube speak, and which, compared to the idiom spoken in the North of the Danube, will be of use to both those of the same origin and for strangers and scholars too.'*

And Boiagi adds, more than that:

> *'Every language is an appearance of the human spirit; the more languages one learns, the*

more sides his mind knows, and hence the
more multilateral he becomes himself.'

Gheorghe Roja

Gheorghe Constantin Roja, a native of Bitolia, published in 1808 the book *Cercetări asupra românilor sau așa-numiților vlahi care locuiesc dincolo de Dunăre* (Research on the Romanians or the so-called Vlachs who live beyond the Danube).

His intention is *'to offer the Aromanians a short description, in particular, so that, as much as possible, they can see what this ancient people was long before and what is it today.'*

In 1809, Gheorghe Roja published another work advocating the use of the Latin alphabet ('the old letters of the Romanians').

Emanuil Gojdu, the great fighter

Figure 38 – *Emanuil Gojdu*

Born in Oradea, Emanuil Gojdu (1802-1870), is the reformer of Hungarian criminal law. A very successful lawyer, prefect of Lugoj and member of the House of Magnates.

The Gojdu family had left Moscopole along with other famous families, Sina and Dumba, reaching Poland. They descend to Hungary, first to Miskolc. A branch of the Gojdu family, Emanuil Gojdu, arrives in Bihor, nowadays Romania.

The house where Emanuil Gojdu was born, relatively modest, only on the ground floor, can still be seen in Oradea next to the Church with Moon, just a few meters from the central square of the city.

Gojdu grew up in Oradea and first studied at the Romanian Orthodox Primary School, then, he goes also to the Catholic high school in Eger, returning to Oradea to study at the Academy of Law, then at the Bratislava Law Academy and the University of Budapest. In 1824, he began his exceptional career as a lawyer and politician.

Gojdu in the office of the well-known lawyer Vitkovics, who was also an acclaimed writer, makes his three-year internship. In his house, the young Gojdu has the opportunity to meet many important writers of the time. This is how he debuted in 1826 with some poems in Hungarian in the *Szépliteratúrai Ajándék* magazine.

At the same time, Gojdu meets some Romanian intellectuals at the house of another Aromanian, Atanasiu Grabovsky, and becomes friends with his nephew, a young law student, Anastasiu Şaguna, who will become the Metropolitan of Transylvania.

Emanuil Gojdu opens his own cabinet and notary cabinet, becoming a respected lawyer, whose interventions become examples for students. The introduction of Hungarian in lieu of Latin in court proceedings in Budapest is his merit.

Gojdu proves to be worthy of his Aromanian tradition, buying in 1832 the house of Wilhelm Sebastian of Budapest, Kogsgasse, known today as the Gojdu Passage (*Gozsdu–udvar*).

In the context of the Revolutions of 1848, Gojdu has a position of devotion and confidence that the Hungarian state will be able to give the laws that correspond to the aspirations of the Romanians, while showing distrust in the Habsburg-style *'divide et impera'*.

After the establishment of the absolutist regime, Emanuil Gojdu took a step back, but accepted his appointment as prefect of Caraş County, becoming member of the House of Magnates, the Upper House of the Hungarian Parliament when the Habsburg Empire restored in 1860 some local legislative autonomy.

Gojdu was a great fighter for the rights of the Romanians in Transylvania, demonstrating, both through his actions and through political statements, that he had not forgotten his Romanian origins.

Gojdu had great hopes for the positive influence of culture, as he says in an appeal for the establishment of a fund for aiding Romanian students at the Faculty of Law in Budapest, in the winter of 1861:

The Romanian nation through culture alone will be able to reach the peak of its greatness, from which our ancestors knew how to instil the respect of the whole world, and who now, with the right of remuneration for the sufferings of 17 centuries, compassionate them.

Only then will she be able to encompass his place among the peoples of Europe, if with diligence, with his power to erase

> *the sad traces left on his face by the time of the ages.* [113]

In the context of the emergence of a stronger national sentiment on a European level, Gojdu consults through letters with other Romanian activists, such as Șaguna or George Barițiu.

One of his most important speeches is that of June 19, 1891, in the *House of Magnates*:

> *I assure the noble Hungarian nation that it is not a good Romanian thinker, not to be perceived by the belief that the divine providence, the God of the peoples of the world, has made the target, for the sake of both the Romanian nation and the Hungarian one, because together they live in an eternal alliance, because only together they have a glorious future, while one against the other, both must perish. [...] Destiny calls these two nations to an eternal alliance.* [114]

Through his tolerant nature, Gojdu was the adept of a rapprochement between the Romanians and the Hungarians, finding a *modus vivendi*.

---

[113] Berenyi, Maria, *Viața și activitatea lui Emanuil Gojdu*, BudaPest, 2002, p.92
[114] *Ibid.*, p.72

> *Let's be cautious, brothers! Wise and wise, and we will not be weary. Let us not forget the sentence of Szecsenyi, the greatest Hungarian, who says: 'You can catch rabbits with the Hungarians - if you go after the pear.' [...] - Wait, brothers, until after the diet! And then judge Gozsdu.*[115]

On November 4, 1869, he signs the famous testament by which he decides that the notorious Gojdu Foundation (Fundația Gojdu) will administer his property[116].

He died a few months later, on 3 February 1870, being mourned in the Kerepesi cemetery in the Hungarian capital. By his will, he left his fortune only to *'that part of the Romanian nation in Hungary and Transylvania, which is part of the Orthodox Eastern Law,'* for scholarships and the help of priests. In this respect, a foundation was established, which operated between 1870 and 1917, giving many scholarships to Romanian students.

In 1952, the Communists nationalize the Gojdu Foundation's property and buildings.

---

[115] *Foaie pentru minte, inimă și literatură*, nr.18, 1861, p.143-146.

[116] 7. întreagă averea mea, [...] o las în întregul ei acelei părți a națiunei române din Ungaria și Transilvania, care se ține de legea răsăriteană ortodoxă. Din lăsămîntul acesta voiesc să se constituie o fundațiune permanentă, care va purta numele "Fundațiunea Gozsdu"; *Testamentul lui Emanuil Gozsdu*, Sibiu, 1899, p.6-22

> *'For Romanians in Hungary, Emanuil Gojdu*
> *represents a point of reference for their*
> *identity. [...]Gojdu was a European, who*
> *has since seen in the spirit of our time.*
> *Immediately after his establishment in*
> *the Hungarian capital, he enters the*
> *highest circles of Hungarian and*
> *Romanian intellectuals, ensuring his*
> *respect and sympathy for all. [...] Gojdu*
> *was a patriot who thought of his own*
> *people, acted as a politician, preferred to*
> *win, to act, to initiate effective*
> *programs.'* [117]

Andrei Șaguna, the Metropolitan

Born in 1808, in Miskolc, Northern Hungary, from Aromanian parents, from Grabova, near Moscopole, Anastasie, Andrei Baron de Șaguna (1808 - 1873), was the first Orthodox Metropolitan of Transylvania in a church whose autocephaly is due to him.

He attends the Greek-Wallachian school in his native town, then goes to study Law in Budapest. Then, Anastasie Șaguna also attends the Orthodox Theology courses in Vârșeț (Vrsac, in today's Serbia, close to the border with Romania). He becomes a monk in 1833 at the Metropolitanate of Carlovit, named Andrew, and for 13 years he will serve in the Orthodox Church of Serbia.

---

[117] Berenyi, Maria, *Viața și activitatea lui Emanuil Gojdu*, Budapest, 2002, p.5

Like Emanuil Gojdu, a good friend since his youth, Andrei Șaguna believed that only by culture can the Romanians be emancipated. That is why, his whole life struggled for the better exchange of education in

Romanian, acting where he could, as a metropolitan, especially in the primary (confessional) primary schools. Under his leadership, the number of these schools reaches almost 800, more than half being set up by him.

What he did for the Romanian church and school was the creation of a modern organization with a pronounced national character.

Figure 39 – *Andrei Șaguna*

ASTRA *(The Transylvanian Association for Literature and Culture of the Romanian People)* owes Șaguna the recognition as a legal person in the State.

Șaguna spent many years in his constant desire to create ASTRA and the first meeting took place on October 23, 1861, with him as the president, an unmatched, visionary administrator and organizer, doubled by a skilled diplomat, a true diplomatic representative of the Romanian people at the Court of Vienna.

In 1846, archimandrite Andrei Șaguna was appointed general vicar of the Bishopric of Sibiu. Șaguna tells the Romanians at his installation on April 18, 1848 as Orthodox Bishop of Sibiu that he wants:

> *'The Transylvanian Romanians from their deep sleep to wake them up and willingly to all that is true, pleasant and good to pull them.'*[118]

Șaguna was a militant for the rights of the Orthodox people and the Romanians in Transylvania, founder of the Romanian Gymnasium in Brasov (1851) and honorary member of the Romanian Academy. Andrei Șaguna is involved in the revolutionary movement of the Romanians, being co-president of the *Blaj National Assembly* in May 1848.

Șaguna founded *Telegraful Român* (the Romanian Telegraph) newspaper on January 3, 1853. He joined the Imperial Senate in Vienna, but was also president of the national conference in Sibiu in 1861, then in 1863 member of the Transylvanian Diet.

In 1864, due to the personality of Șaguna, the Bishopric of Sibiu was raised for the second time to the rank of Metropolitan, which was abolished in 1701, with its own autonomy, detaching itself from the Serbian Orthodox Metropolitan Church of Carlovit.

Seeing the drama of the Romanian school infrastructure, Șaguna has in 1852 the idea of obliging each of his parishes to raise a school, the Eparchial Publishing House being the one to provide the only textbooks to be used in these schools.

---

[118] Acu, Dumitru, Prof.univ.dr., Andrei Șaguna, *Aociațiunea ASTRA și lupta pentru unitate națională*, http://www.tribuna.ro/stiri/cultura/andrei-saguna-asociatiunea-astra-si-lupta-pentru-unitate-nationala-44738.html

The attention given by Șaguna to the public schools is known, with no less than 25 textbooks prepared with his priestly support from his diocese. He goes on and orders in 1870 priests to hold special courses for the illiterates.

Unfortunately, he was not able to accomplish everything he wanted and he founded a single high school in Brașov in 1850, who still has his name today. Pupils deprived of material means have provided aid from his own income or through the special foundations.

One of these foundations, created by Emanuil Gojdu at the suggestion of Șaguna, *the Gojdu Foundation*, functioned very well between 1871 and 1917. In 1864, Bishop Andrei Șaguna is named Metropolitan of Transylvania. Under his guidance, almost 800 schools are set up in the Archdiocese of Sibiu, directly under the guidance of the Orthodox Church.

'He is part of the family of those forefathers for whom the sutana has never been a sign and a call to religious exclusivity but, on the contrary, gave them at that time the necessary authority to represent and call the Romanian people to light, freedom and independence.

They were not one of those whose bells only called for church and office, but also for school and for the collection of words for the press of books of all kinds, of literature and science, of mathematics and of engineering,'[119] another great

---

[119] Vlad, Valentin I., Acad., *ASTRA și cultura poporului român*, p.1
www.acad.ro/com2011/doc/ASTRAacadVlad.doc

Metropolitan of Transylvania, Dr.
Antonie Plămădeală, says of him.

Panu and Romania's strategy for the Balkans

Anastasie Panu (1810-1867), born in 1810 in Iași, as a son of Panaiotache Panu, a Macedonian Aromanian, was a politician and lawyer.

Panu took an active part in the Revolution of 1848, becoming Minister of Justice during the reign of Grigore Alexandru Ghica in 1852.

Later on, during the reign of Alexandru Ioan Cuza, Panu was Chairman of the Council of Ministers and Minister of the Interior, and he outlined a clear program of how the United Principalities, and then Romania, should treat the situation in the Balkan Peninsula.

Lefkada, a great poet of modern Greece

Aristotelis Valaoritis Lefkada (1824-1879), Greek national poet and politician, grandfather of Nanos Valaoritis, one of Greece's most respected writers.

He was born to Lefkada, his father having a descendant from Epirus. He studied law in France and Italy.

Member of Parliament in Athens, is known for his amazing remarks.

Ion Ghica, academician and Prime Minister

Ion Ghica (1816-1897), born in Bucharest in 1816, is a distinguished personality of Aromanian origin in the second half of the nineteenth century, academician, author, diplomat, mathematician, politician and Romanian pedagogue.

He was Prime Minister of Romania twice, between 1866 and 1867, and between 1870 and 1871, as well as President of the Romanian Academy four times between 1876 and 1895.

Son of Dimitrie Ghica, educated in Bucharest and Western Europe, studied engineering and mathematics in Paris between 1837 and 1840. He returned to his country and was part of the 1848 conspiracy that wanted a union of Wallachia and Moldavia under a local prince, Mihail Sturdza.

Together with Nicolae Bălcescu and Christian Tell, he founded the Masonic Society *'The Brotherhood'* *(Frăția)*, the engine of the Revolution of 1848.

Mărgărit, apostle of Romanian spirituality

Apostol Mărgărit (1832-1903), born in Abela, is a true apostle of Romanian spirituality, a well-known Aromanian teacher, a correspondent member of the Romanian Academy.

Apostol Mărgărit is mostly known as the national leader of the Aromanians in the Ottoman Empire, but also as a teacher and school inspector for Macedonia.He becomes a teacher in Vlacho-Clisura at the age of 30, teaching children the Greek and Aromanian.

As a militant for the gathering of diplomatic relations between Romania and the Ottoman Empire, Apostol Mărgărit said:

'Our first interest, the Aromanians, is the salvation of the Ottoman Empire. We do not hope to join hands with our Romanian brothers: we are separated from them by princes and kingdoms...

An Oriental Crisis would give us into the hands of Serbs, Greeks or Bulgarians, Christian and civilized peoples, who, already taking into account the community of religion, would like [...] to shut our schools, scatter our communities'.

The Greeks accused him of being a foreign agent, more specifically an Austrian or Catholic agent, being persecuted by the Patriarch of Constantinople, having to leave school; he tried to give private Aromanian lessons.

He escaped alive after several assassination attempts against him, being stabbed in Thessaloniki, thrown twice in the Vardar River, shot somewhere in the mountains of Ohrida. He gets to jail, but escapes to Bucharest, where he enjoys the support of King Carol.

The Independence War of 1877-1878 brings a radical change in his life. The Ottoman Empire accepts him back as a school inspector at the Romanian schools in the Empire's territory.

This position allows Mărgărit to set up several Aromanian schools in Macedonia and Albania, some of them alongside French priest *Jean-Claude Faveyrial.* Member of the Romanian Academy in 1889.

Ioan D. Caragiani and the Romanian Academy

Ioan D. Caragiani (1841-1921) is one of the founding members of the Romanian Academy. Member of the *Junimea Society*, he was born in Abela.

He studied ancient Greek with the famous Apostol Mărgărit and later on, he became a student of the University of Athens.

Caragiani wanted to create an Aromanian state, located in Pindus. He arrives in Bucharest with an Aromanian caravan and becomes a professor at the University of Iasi.

Figure 40 – *Ioan D. Caragiani and the founding members of the Romanian Academy*

Ioan Caragiani's first writing on Aromanians is *Românii din Macedonia și poezia lor poporală* (The

Romanians of Macedonia and their Folk Poetry), published in 'Literary Conversations,' 1869.

In addition to presenting their Aromanians and their origins, we find here an estimate of the number of Aromanians, namely 1,450,000, but also an appeal to help the Southern Danube brothers, subjected to a powerful denationalization process.

Caragiani supports the idea that the language spoken in the Balkan Peninsula is a dialect of the Romanian language. About their origin, Caragiani had a simple theory: some passed the Danube and the others came from Italy sometime in the past.

Belimace, author of the Aromanian Hymn

Constantin Belimace (1848-1934) was born in Moloviste, Macedonia. After his primary studies in his

native village, he goes to the Serbian school in Belgrade.

Belimace is one of the founders of the Aromanian cult literature, with a warm, popular popular troubadour style.

His work *Dimândarea pârintească* (Blessing from Parents) becomes so popular in a very short time that it is still considered as the true Aromanian hymn.

Figure 41 –
Constantin
Belimace

In 1873 he moves to Bucharest, where he opens a pub which becomes the favorite meeting place for the Aromanian intellectuals.

Spyridon Lambros, Prime Minister

Figure 42 –
Spyridon Lambros

Spyridon Lambros (1851-1919), Prime Minister of Greece, was born in Corfu and studied history in London, Paris and Vienna.

From 1890, we find him at the University of Athens, where he was also the rector.

In 1903, Lambros started an academic movement called *Neos Hellenomnemon*, studying the evolution of the Greek-speaking world in the Byzantine and Ottoman era.

In October 1916, he accepted to form the Government.

Kostas Krystallis, Greek writer and poet

Kostas Krystallis (1868-1894), writer and poet, representative of Greek pastoral literature, born in the Aromanian village Seracu, Epirus, is exiled from the Ottoman Empire for 25 years due to his patriotic writings, reaching Greece.

In 1893 he remained without any job, but won the lottery and could thus afford the publication of his writings.

Asdreni, important Albanian poet

Aleksandër Stavre Drenova (1872-1947), considered to be one of the most famous Albanian poets, was born on April 11, 1872, in Drenova, about five kilometers from Corcea, in today's Albania.

He composed the lyrics of the Albanian hymn. The hymn music, more precisely *'On Our Flag is Written Union,'* is written by the Romanian composer Ciprian Porumbescu. His pen name is *Asdreni.*

His poem 'Song of union', appearing in the volume 'Dreams and tears', 1912), is adapted to the famous *'Hora of the Union'* of Vasile Alecsandri.

In 1885, he leaves for Bucharest, where he joins his two brothers.

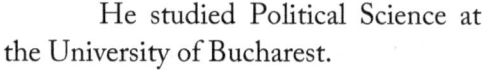

He studied Political Science at the University of Bucharest.

The first volume of poems, entitled *Sun Rays*, dedicated to the great Albanian hero Skandreberg, appears in 1904.

Figure 43 – *Aleksandër Stavre Drenova*

He made a visit to Albania in 1937, hoping to get a state pension. Without success, he died in Bucharest in 1947, poor.

The Manachea brothers (Manaki)

The *Manachea brothers, Ienache* (1878-1954) and *Miltiade* (1882-1964), known as *the Manaki brothers,* precursors of the cinema in the Balkan Peninsula. Manachea Ienache is on the list of graduates of the high school in Bitolia in 1896-1897 with the occupation of the photographer.

Figure 44 - *The Manaki brothers*
Manaki brothers photo collection

In their first film, which is also the first film in the Balkans, their 114-year old Despina grandmother appears weaving. In 1904, the two brothers moved from Abela to Bitolia and opened a photographic workshop.

They attend, at the invitation of King Carol I, in 1906 at the World Photography Exhibition in Sinaia, where they receive a gold medal for their photographic collection.

They become official photographers of the Ottoman Emperor and King Alexander Yaroslavl Karageorgevich. In those times with important historical events, they record these moments on the film, creating a real historical archive.

Now, the *National Archives of the Republic of Macedonia*, holds more than 17,000 photos and over 2,000 meters of film from the Manaki brothers.

They are honored with an international film festival each year in their hometown of Bitolia.

Figure 45 - *Poster from a*
Manaki brothers film
Manaki brothers photo collection

GREAT FAMILIES

Mocioni family

Figure 46 - *Coat of arms - Mocioni*

Figure 47 - Coat of arms - Mocioni de Foeni

Mocioni, appointed also *Mucioni*, Magyarized *Mocsonyi*, is an Aromanian family emigrated to Hungary and settled in Banat, actively involved in promoting and achieving political rights of Romanians in Hungary. In Moscopole, they were nicknamed *Moceanu* or *Mucianā*, that comes from *Mocian*, a name given to the Aromanians of Aspropotam of Thessaly.

They were related to Hungarian, Serbian and German families, but they kept an unwavering orthodox conservatism and have kept away categorically from the magyarization current .

Mocioni family tradition of Moscopole says that two brothers fought for the Christian powers, one fell in the battle of Zenta (1696), the other in the battle for releasing Timisoara. The fact is that the *priest Mocioni Constantine (Constantinus Motsonyi)* is mentioned in Hungary, which has five sons, large traders in Budapest. They soon managed to gather great wealth and even to be ennobled.

Indisputable historical fact, resulting in a genealogical tree at the end of the eighteenth century, however, states that 'a Constantinus Motsonyi', Greek presbyter ('Presbyter or Popa Gr. N. Unitus»), came in 1747 in Hungary from Macedonia ('Macedonia ex Anno 1747 in Hungary advena') and died at the age of 110 years in Pest

('Pestni suae anno mortuus Aetatis 110').

Father Constantin Mocioni had - genealogy informs us - five sons ('5 Fili habens'), of whom only two are mentioned and named: Andrew and Mihaiu who, '[and] beside the name of Motsonyi» they add the name « Popovits », because – it is noted in genealogy - their father was a priest (' Popovits ideo quod dicti Popa fuer AVUS '). The name 'Popovits',

> *'Popovics', 'Popovich' (Popovici),*
> *together with the name «Motsonyi',*
> *'Mocsonyi», passed in their ennoblement*
> *diplomas.* [120]

Out of the five sons, Andrew founded the *Foeni* line, after obtaining in 1780 the land from Foeni, now at very short distance from the border with Serbia. Michael, who also had five sons, founded the *Mocioni* line.

The land purchased by him is at Tokay. Andrei Mocioni gets a violent death, being shot through the window by a person remaining even today unknown in 1782 in the house of his domain administrator from Foeni.

Proving that they have the typical Aromanian entrepreneurship, Mocioni family members come to be an important political, religious, and cultural factor, being influenced also by the powerful Romanian cultural trend manifested in the Hungarian capital since the years prior to the revolutions of 1848.

Thus, besides preserving a certain level of material wealth, they manage to increase it from generation to generation, giving them the status of independence of opinion, not only prestige.

Somehow different from other Aromanians families, despite all its kinship relations with the families of Hungarian, Serbian and German origin, Mocioni family resisted to inherent temptations in removal of traditions, as to the magyarization current.

---

[120] Dr. Botiş, Teodor, Monografia familiei Mocioni, FUNDAŢIA PENTRU LITERATURĂ ŞI ARTĂ „REGELE CAROL II", Bucureşti, 1939, p.18

In 1865, when Count Zichy, auditorium Chancellor of Hungary, communicates - through a family friend - the intention to propose Emperor Francis Joseph to raising the Mocioni family members to the rank of counts, which would have given them the title and dignity of a member of law of 'House of the Magnates' (The Hungarian Senate), Andrei Mocioni replies:

'Tell him to give me back at least peace, unless he cannot repay the confidence kidnapped by those on top'', and his brother Antony declares

'Those on top think we are wicked or insane and do not know that I would be ashamed to go out among the people with the title of coont (Graf), as long as the miserable situation of the Romanian nation lasts today.[121]

From among the members of the Mocioni family were distinguished throughout time:

Petre Mocioni (1804 - 1858) - MP,

Andrei Mocioni (1812 - 1880) - politician,

Anton Mocioni (1816 - 1890) - officer, MP,

George Mocioni (1823 - 1901) – MP,

Alexandru Mocioni (1841 - 1909) - writer, journalist, politician, MP, composer,

Zeno Mocioni (1842 - 1905) - lawyer, politician,

---

[121] *Ibid.,* p.29-30

Eugen Mocioni (1844 - 1901) - MP,

Ionel Mocioni (1893 - 1930) - lawyer, politician, member of the Great National Council, deputy prefect of Severin,

Ion Mocsony Styrcea (1909 - 1992) - Marshal of the Royal Court.

The Sina Family

Sina family motto:
Servare intaminatum

Gheorghe Simeon Sina

Figure 48 - *Gheorghe Simeon Sina*

George Simeon Sina is the most famous member of his family, one of the most important families coming from Moscopole.

The Sina family greatly contributed to the support and assertion of the Romanian cultural movement in Hungary, but also to the support of the Orthodox Church there.

His father, *Simeon Sina*, held at that time the second major bank after that of Rothschild's, considerably enriching the import and trade of tobacco, growing in parallel, significantly, the trade between Austria and Turkey.

Gh. S. Sina lead a wise policy of purchasing several properties in Hungary, Transylvania, Croatia and Țara Românească (Valahia), making the Sina family one of the richest families in the nineteenth century Europe. Following the natural course of his social ascent, he becomes a baron in 1818, after buying some estates from Hodos and Kizdia.

Figure 49 - *The Sina Palais on a bigger map of Vienna,* by *Carl Graf Vasquez*

wien.gv.at/kultur/kulturgut/plaene/karten/images/vasquez-gesamtwien1.jpg

Thanks to the close relationships he have had in St. Petersburg, Sina's Bank enjoys the confidence of the Russian aristocracy, as many deposits came from Czarist Empire. Perhaps this explains the Czarist propaganda financing in the Balkans.

George Simeon Sina has been side by side with great reformer *Istvan Szechenyi*, not only in oral requests, but in also especially in financing the establishment of important institutions of the Hungarian state.

Sina has a great contribution to the construction of the famous 'Chain Bridge' over the Danube. Simeon George Sina made important donations for the establishment of state institutions such as *the Hungarian company of insurance*, *Agricultural Credit of Hungary*, *Commercial Academy*, the *Hungarian National Theatre* and especially *the Hungarian Academy*.

Here are the official qualifications of Baron Sina:

'Son Excellence Monsieur le Baron Simon Sina de Hodos et Kizdia, Envoyé Extraordinaire et Ministre Plénipotentiaire du Roi Othon Ier près les Cours d'Autriche, de Russie et de Bavière, Grand-Croix de l'Ordre Royal du Sauveur de Grèce, de l'Aigle Rouge de Prusse, première classe, de l'Ordre Royal de Saint-Michel de Bavière et de l'Ordre Grand Ducal d'Oldenburg pour le Mérite, Commandeur de l'Ordre Impérial de Léopold d'Autriche et de l'Ordre Impérial de la Légion d'honneur de France, et propriétaire des Ordres Impériaux Ottomans du Nishan Iftikhar et du Médjidié'.

Mihai Eminescu said in 1881 in the Romanian newspaper *The Time*:

> '*Sina established the Greek community in Vienna and raised funds for the Greek insurrection, a thing that required courage at the time when Metternich lived, who called it 'the most undignified rebellion ever brightened by the sun.' What did his son, the late Greek ambassador Sina, in heightening the arts and sciences in Athens by buildings and generous donations, is still in the memory of all.'*

In a presentation of the Dumba family, Wittmann Hugo writes in 1915 about those two great Aromanian families of Vienna, Sina and Dumba:

> '*Sina family has come even before Dumba in Austria. The old Sina, grandfather of Baron Sina, who died in 1876, was the predecessor, followed [...] by the father of our Nicholas Dumba, founder of the great trade, the owner of several factories and spinning factories, the builder of the familiar palace from the corner Zetlitz street and the boulevard (Ring) of Emperor Wilhelm, Ottoman general consul, otherwise called, simply, Sterio (Sterie) Dumba, not baron, [...] but simply, bourgeois, Sterie Dumba. Just*

*like old Sina,he also came as a poor kid
in a carriage full of cotton to Vienna.'[122]*

The Dumba Family

Figure 50 - Nikolaus Dumba in his office in Dumbas Palace,
Vienna

Nikolaus Dumba, the most popular member of
the Dumba family (July 24, 1830, Vienna, March 23,
1900, Budapest) was a great businessman, but also a

[122] TRANSSILVANICA, Biblioteca digitală, Biblioteca Centrală
Universitară "Lucian Blaga" Cluj-Napoca,
http://documente.bcucluj.ro/web/bibdigit/periodice/transilvania/1925/
BCUCLUJ_FP_279996_1925_056_005_006.pdf, p.11

prominent liberal politician. He is known as patron of the arts, collector and promoter of musical life in Vienna. A much detailed description is set out at the end of this book, in the Annex.

Stergios (Sterie) Dumbas was the first in Dumba line who emigrated to Vienna in 1817 from Vlasti, a village in Northern Greece today. Here he was known as a great merchant, who took his chances to change the European policies in cotton, opening, at his own risk, a new trade route to the Orient, by bringing Macedonian cotton. Ana he succeded, because it seems that this clever maneuver was the basis of his wealth, combined with increasing exports to the Ottoman Empire.

Once arrived in Vienna, the Dumba family quickly became one of the richest merchant families of the beginning of the nineteenth century. However, the Dumba family members did not forget their origins, despite their deliberate confusion with the Greeks, therefore supporting the Romanians' activities from the capital of the Habsburg Empire, Vienna.

The Dumba family was a big supporter of the student society 'România Jună' in Vienna and for many years 'imperial council member Nic. Dumba' was the honorary patron festivities and balls.

'Papa Dumba' helped many Romanian students and was the protector of the society 'România Jună' at the celebration of 25 years (1896).[123]

---

[123] *Ibid.*, p.11-2

Just like the Sina family, the Dumba family remained in the Viennese memory for a long time. In 1915, a story called *'Die Dumbas'* (the Dumba family), appears in the journal *Neue Freie Presse* in Vienna. Now, this is a publication that did not love the Romanians too much, therefore the article written by the critic and esthete Hugo Wittmann would be a very interesting article from the point of view of Aromanians.

'The Aromanians... they say they are very much loved in the Balkans.

They are good in trade, gifted with a merchant spirit, always business-oriented, and many of them have shown what can bring diligence and perseverance at the court of the Sultan, if supported by reason, cleverness and a certain cunning life art. About this, we, the ones from Vienna, know how to tell, in addition, a more beautiful story.

[...] We can exemplify with two brilliant pieces of evidence that even from a Macedonian living so far away (from Vienna) we can obtain a good Viennese, an authentic, real one, an excellent Viennese – we are referring to the Sina and Dumba families.

It is in particular the last name [...] 'Dumba', whenever we used to mention his name, or 'der Nickel Dumba', even the great men of stone and bronze: Schubert,

> *Mozart, Beethoven, Makart, would*
> *tremble on their pedestals on hearing it,*
> *because this is the man they had to*
> *thank for the foundation of their*
> *monuments.*[124]

Nikolaus Dumba was, *inter alia*, Deputy Governor of the Commercial Bank Vienna, Member of Austrian Parliament and an active participant in the cultural life of Vienna, advisor, promoter and patron of many arts and cultural initiatives. By his will, he made a donation of 200 Schubert autographs to the city, creating one of the largest collections of music in the world, *Vienna City Library*.

As a patron of arts, he collected and promoted the contemporary art of his time and helped artists, being also a friend of *Hans Makart, Gustav Klimt* and *Carl Kundmann*. As a deputy, he suggested the construction of numerous monuments of composers. Nikolaus Dumba was in contact with *Johannes Brahms, Richard Wagner* and *Johann Strauss*, but he preferred *Franz Schubert*.

As a splash of color, Johann Strauss (son) wrote his famous *Blue Danube* as a waltz for choir, back when *Vienna Choral Association for men* had Nikolaus Dumba as its conductor (1865-1872).

Thus, Nikolaus Dumba is the founder of the University of Athens, for which he donated the funds necessary for interior decorations and furnishings. In Serres, where Sterio Dumba learned the skill of a silversmith and became initiated in the cotton and tabacco industries, he built the orphanage, which is now a kindergarten. In addition to this, he was actively involved in building vocational schools there.

---

[124] *Ibid.*, p.13

More squares and streets of Austria today are called Dumbas. *Dumbas Street* in Vienna (*Dumbastraße*) starts from the *Imperial Hotel Ring* and ends at the *Musikverein*.

Of the political positions held by Nikolaus Dumba, we recall here that, between 1870 and 1896, he was a member of *Niederösterreichischen Landtages* (Parliament) and in 1885 he was appointed by the Emperor as a lifetime member of the *Herrenhaus*, the Upper House of the Austrian Reichsrat, Imperial Council.

For his work, Nikolaus Dumba was honored, among others, with several medals and decorations to several countries:

> • *Member (1870) and honorary member of the Academy of Fine Arts Vienna (1880)*

> • *Honorary Member of the Society of Friends of Music in Vienna (Gesellschaft der Musikfreunde in Wien. In short: Musikverein) (1877);*

> • *Honorary Citizen of Vienna (July 25th 1890);*

> • *Cross of Knight of the Order Franz-Joseph;*

> • *Mecidiye Medal Class II of the Ottoman Empire;*

> • *Cross the Grand Commander of Greece Tagma tou Sotiros:*

> • *Grand Officer of the Romanian Order of the Crown of Romania.*

His son Konstantin (Theodor), Count of Dumba, a diplomat, was the Austro-Hungarian ambassador to London, St. Petersburg, Rome, Paris, among others, but also to Bucharest.

As an Ambassador to the United States during World War I, he was expelled following a much publicized diplomatic scandal, accused of spying and sabotage attempt of the US arms industry.

Konstantin, Count of Dumba, is, the last ambassador to the United States of the dying Austro-Hungarian Empire.

THE VANISHING AROMANIAN?

Romania and the vanishing Aromanian

For the last 150 years, Romanians did not want or could not understand who are the Aromanians. The policy of establishing schools before 1900 in the Balkans to teach the Romanian language, instead of the Aromanian one, is something that today we would call assimilation.

If only somebody in charge at the beginning of the century had known how things would unfold... Nicolae Iorga said this in 1928:

'The new rising Macedonian affair, [...] – might have been better to talk about a culture delivered in the Aromanian dialect, for it would be more useful –, not touching the vital interests of Romania, so someone could look quietly how things are developing.

Romania was to be satisfied with recognition (May 1905) of the Aromanian nationality from the Turks. '[125]

The Aromanians, as a community, may have their momentary affinities, perhaps, they could be divided, but not blamed for this typically human weakness. Not even the Aromanian politicians could understand the need for a smart, coherent strategy, detached from political struggles.

Yes, we cannot turn back time, but we can understand more and act before it's too late...

---

[125] Iorga, Nicolae, *Istoria poporului românesc*, trad. din lb. germ. de O.E. Ionescu, vol. IV, partea a II-a, Ed. Casei Şcoalelor, 1928, p.72

Aromanians and Romanians today

From the very beginning, I knew this would be a sensitive subject, but I could not just watch and do nothing, as a silent witness to the disappearance of the Aromanians' identity. I tried to understand more, so I went to Moscopole several times, I visited communities in Macedonia, Albania, and Greece. I wrote three books in which I tried to explain once again why this *Aromanian affair* exists and what is there for the Greeks and for the Romanians.

This is apparently the most delicate problem: how to explain to a typical Romanian that an Aromanian is different enough, that he should not be considered a Romanian.

Now, more than ever, the interests in the assimilation of outside groups in a given country, as being of the same origin, are stronger than ever. The situation is even more acute when it comes to member states of the European Union, which know that having more citizens offers them more power at the common European table.

No matter how much and where this numerical growth comes from, because, you know, as Macchiavelli said: *'For although the act condemns the doer, the end may justify him[126]...'*

The Aromanians, once again in their history, are trapped between various state interests.

---

[126] Macchiavelli, *Discourses*: I, 9

The new Millenium Aromanian

'Aromanian (also known as Macedo-Romanian, or Vlach) is spoken by half a million people, mainly in parts of Northern Greece. Albania, the Former Yugoslav Republic of Macedonia, and Bulgaria.

It constitutes the second most populous variety of the Daco-Romance branch of the Romance languages, the others being: Daco-Romanian (with about 25 million speakers, mainly in Romania and Republic of Moldova).

Istro-Romanian (spoken by at most 1500 souls in the Istrian peninsula of Croatia), and Megleno-Romanian (with some 5000 speakers principally in areas of Greece and FYR Macedonia to the North of the Gulf of Salonika),[127] says Martin Maiden.

Nowadays, Aromanian is no longer the Middle Ages' Vlach, nor the Aromanian banished from his own house and chased by armed hordes beyond Vienna, far into Poland. Nor is nowadays the Aromanian a EU citizen with full rights, still living in a world still divided into several small Balkan countries.

---

[127] Maiden, Martin, *"Perfect pedigree. The ancestry of the Aromanian conditional"*, in Ashdowne, Richard and Finbow, Thomas (eds.), Oxford Working Papers in Linguistics, Philology and Phonetics, 9: 83-98, 2004.

And the danger of losing his identity still stands above his head like the sword of Damocles and here is the most serious one: *the danger of assimilation.*

> *'Today, after two Balkan wars and two world wars,*
> *with serious consequences in all areas, the*
> *Aromanians' situation is dramatic in the*
> *Balkans.*
>
> *Failing to impose countries in the region, very soon,*
> *schools or classes taught in the Aromanian,*
> *the language that withstood those regions*
> *more than two thousand years will go off*
> *before our eyes.'*[128]

The above words, said by the stylish erudite Neagu Djuvara, a man who has seen so many during his long life, are more than just a warning.

What chances of survival would the Aromanian identity have in this modern world of globalization, away from the traditional model of the Aromanian community? I wrote this book with the belief that their chances must exist, that this attempt is not in vain.

In a world without borders, the Aromanian, who has always been a European citizen, feels at home everywhere.

Maybe the traditional Aromanian is in danger, but he is not the same as the Aromanian of the nineteenth century. His values, even inclined to the traditional style of life, could be adapted to our modern times.

The new Aromanian, which is a combination of traditional and modern, must rise from the ashes of the old, such as Phoenix, not only as the product of his own mind, but also as his creative soul of light.

---

[128] Djuvara, Neagu, O scurtă istorie a românilor povestită celor tineri, Ed. Humanitas, București, 2010, p.259

As direct descendants of Alexander Makedon, old Macedonians, with tradition, language, history, and flag, Aromanians will not be Romanian or Greek, or anything else. They will remain what they have been for 4,000 years: Aromanians.

And those who understands this, in time, will be winners!

Still, no one will lose. This is a win-win situation.

Perhaps the future reserves for us something special and we have not learned that yet...

ANNEXES

Shared admixture events in Eastern Europe in the first Millenium

Inferred events were broadly similar among 8 groups from Eastern Europe, so we re-analysed them, excluding each other as donors (EastEurope1 analysis; see Note S7.6).

The two most easterly groups, the Russians and Chuvash, show similar signals with inferred admixture between two groups, one with ancestry related to Northeast Asians and one related to Europeans (Table S16). There is evidence in both groups (p<0.05) that this occurred at more than one time (Figure 2D), with very ancient East Asian ancestry prior to 500BC, in the case of Russia at least, and a recent event, consistent with approximately Mongol-empire era admixture, together contributing ~10% of DNA in Russians and ~35% in the Chuvash.

Six other eastern European populations show highly shared events (Figure 3). None of these populations shows evidence of multiple admixture times (p>0.5), but all six independently show evidence for admixture between more than two groups (p<0.02). For example in Bulgarian genomes, DNA chunks more closely related to those carried by present-day Greek, Norwegian and Oroqen people tend to be separated in the genome so that each pair of curves shows a dip at short genetic distances (Figure 3), implying these segments must occur on three different ancestry backgrounds. The clearest admixture signal in each population predates the Mongol empire but involves the minority source group having at least some ancestry related to Northeast Asians (e.g. the Oroqen, Mongola and Yakut), with approximately 2-4% of these groups' total ancestry proportion linking directly to East Asia, lowest in our Polish sample and highest in Hungarians.

A second signal in each group involves admixture between distinct European groups – one more Southerly (sharing more DNA segments with e.g. Greece and W. Asia) and one more Northerly (sharing more DNA segments with e.g. North and North-west Europe) – at approximately the same time as the east Asian admixture, and with each group inferred as contributing relatively similar amounts of DNA (though the fraction is uncertain). Inferred admixture dates fall within a relatively tight range (440-1080CE), earliest in Poland, with groups more distant from Poland generally having more recent dates.

These results are consistent with our detecting a genetic legacy from invasions of peoples from the Asian steppes (e.g. the Huns, Magyar and Turkic Bulgars) during the first millennium CE (1, 2), and affecting all six groups. We speculate that the second event seen in our six Eastern Europe populations between sources more similar to Northern-Europeans and Southern-Europeans and West Asians respectively may correspond to the expansion of Slavic language speaking groups (commonly referred to as the Slavic expansion) across this region at a similar time, perhaps related to displacement caused by the Eurasian steppe invaders (2, 3).

This would imply the more Northerly donor population might represent Slavic speaking migrants. To test consistency with this, we repainted these populations now adding the Polish as a single Slavic donor (EastEuropeII analysis, see Note S7.6) and, in doing so, they largely replaced the original North-European component (Figure S21), although we note that two nearby populations, Belarus and Lithuania, are equally often (relative to the Polish) inferred as sources in our original analysis (Table S12).

Outside these six populations, an admixture event at the same time (910CE, 95% CI:720-1140CE) is seen in the Southerly neighbouring Greeks, between sources represented by multiple neighbouring Mediterranean peoples (63%) and the Polish (37%), suggesting a strong and early impact of the Slavic expansions in Greece, a subject of recent debate (4). These shared signals we find across East European groups could explain a recent observation of an excess of IBD sharing among similar groups, including Greece, that was dated to a wide range between 1,000 and 2,000 years ago (4).

References

1. C. Atwood P., 'Encyclopedia of Mongolia and the Mongol Empire', Facts on File, Inc, New York, USA, 2004

2. C. I. Beckwith, 'Empires of the Silk Road: A History of Central Eurasia from the Bronze Age to the Present', Princeton University Press, Princeton, US, 2006

3. D. W. Anthony, 'The Horse The Wheel and Language: How Bronze-Age Riders from the Eurasian Steppes shaped the Modern World', Princeton University Press, Princeton, US, 2007

4. P. Ralph, G. Coop, 'The Geography of Recent Genetic Ancestry across Europe', PLoS Biol 11, e1001555 (2013).

A genetic atlas of human admixture history
Companion website for 'A genetic atlas of human admixture history', Hellenthal et al, Science (2014). http://admixturemap.paintmychromosomes.com/

*** References from Diamandi-Aminceanu, *Românii din peninsula balcanică* (Romanians from the Balkan Peninsula), Bucharest, p. 12, to authors and works known more or less as being related to the Aromanian origin issue, refer respectively to:

3. Le Beau în Hist. du Bas Empire; Hammer in Hist. de l'Empire Ottoman; Jirccek in Geschichte der Bulgaren şi in Die Romanen in den Städten Daimatiens; Cousinery in Voyage dans la Macédoine.

4. Ami Bouiê in La Turquie d'Europe; Poujade in Chretiens et Turcs; Jules Blancard in Épire et Thessalie; Ubicini in Les Origines de l'histoire roumaine; El. Reclus in Géographie Universelle; Joncquière in Hist. De l'Empire Ottoman; Leake in Researches in Greece; Encylopedia Britanica; Sulzer in Geschichte des Transalpinischen Dacien; Engel in Geschichte der Moldau und Valachien; Weigand in Die Aromunen; Miklosich în Slavische Elemente in Rumunischen.

5) Heuzey in Le Mont Olympe; Thunmann in Untersuchungen über die Geschichte der öftliehen Europäischen Völker; Falmeraytr in Fragmente aus dem Orient; Hahn in Reise von Belgrad nach Salonik; Kiepert in Ethnographie von Epirus; Hertzberg in Geschichte der Byzantiner und des Osraanischen Reiches; Jung in Römer und Romanen in den Donauländern; Tomasch k in Zeitschrift für die Ocst. Gymn. din 1877; Buruy in History of tbe Later Roman Empire; Finiay in History oi Greece; Lamouche in La Péninsule Balkanique.

6) Rössler in Rumänische Studien; Schafarik in Slavische Alterthümer,

278

Die Aromunen. Ethnographisch-philologisch-historische Untersuchungen: Vorrede.

Wie ich schon in dem zweiten, vor einem Jahre erschienenen Bande gesagt habe, bezieht sich der Name 'Aromunen' auf das walachisch redende Volk der südwestliehen Balkanhalbinsel, das mit Makedo-Walachen, Südrumanen, Pindus-Walachen, transdanubische Walachen, Kutsowlachen, Zinzaren etc. bezeichnet wird.

Alle diese Namen sind unzutreffend in Beziehung auf Verbreitung und Herkunft, oder Spottnamen, wie die beiden letzten. Da man sieh doch schließlich einmal über einen Namen einigen muss, habe ich denjenigen vorgeschlagen, den das Volk in allen seinen Gebieten sich selbst zulegt, nämlich 'Aromunen', welches die deutsche Wiedergabe für 'Arămâni' ist.

Nicht entfernt ist es mir in den Sinn gekommen, diesen Namen zu erfinden, oder gar in Anlehnung an Miklosich's Form 'Rumunen' zu bilden, wie mir dies Gustav Meyer in seiner Kritik in den indogermanischen Forschungen unterlegt hat. Für die Daco-Rumänen hat man ja eine allgemein gebräuchliche Form: Rumänen, da war Miklosich's 'Rumunen' überflüssig, aber für die 'Aromunen' besteht keine einheitliche Form, daher ist eine solche notig.

Was war also natürlicher, als ihnen ihre eigene Bezeichnung zu belassen, allerdings unter der deutschen Form; die Rumänen mögen Arămâni oder Armâni schreiben, für sie geht das an, für Deutsche aber nicht, weil ihnen die Laute ă und î unbekannt sind. Daß man nicht die von Hugo Schuchardt und anderen bevorzugte Form Südrumänen

wählen kann, habe ich Seite 296 dargelegt; damit kann man
die südlich der Donau wohnenden Daco-Rumänen
bezeichnen, die sogar abzüglich der Rumänen der
Dobrudscha zahlreicher sind, als die Aromunen.

Wenn jemand einen passenderen Namen
vorgeschlagen hätte, würde ich ihn gern angenommen haben,
da das nicht geschehen ist, bleibe ich bei dem nicht von mir
erfundenen, sondern allgemein bei dem Volke verbreiteten.

WEIGAND, GUSTAV, Die aromunen. Ethnographisch-philologisch-historische
    untersuchungen, Leipzig, 1895.

The same text translated by me in English

Preface.

As I said in the second volume, published a year ago,
the name 'Aromunen' refers to the Wallachian-speaking people
of the southwestern Balkan Peninsula, referred to as Makedo-
Wallachians, South Romanians, Pindus-Walachians,
Transdanubian Wallachians, Kutsowlachen, Zinzaren etc. All
these names are incorrect in relation to distribution and origin,
or mocking names, like the last two.

After all, since one finally has to agree on a name, I
have proposed the one that the people in all its fields gain for
themselves, namely, 'Aromunen', which is the German
rendition for 'Arămăni'. It did not occur to me to invent this
name, or even to build it on the basis of Miklosich's 'Rumunen'
form, as Gustav Meyer underlined this in his critique of Indo-
European research.

For the Daco Rumanians one has a common form:
Romanians, Miklosich's 'Rumunen' was superfluous, but for

the 'Aromunen' there is no uniform form, so this is necessary. So what was more natural than to leave their own name, but under the German form? the Romanians may write Arămăni or Armăni, it is for them, but not for Germans, because the sounds ā and ī are unknown to them.

I have set out on page 296 that one can not choose the form preferred by Hugo Schuchardt and others. this is the Daco Romanians living south of the Danube, who are even more numerous than the Aromanians, minus the Dobrogea Romanians. If someone had suggested a more suitable name, I would have liked to have accepted it, since this has not happened, I remain with the one not invented by me, but generally spread among the people.

Voyage de la Grèce, Pouqueville

Excerpt regarding the origin of Vlachs, in orig., Fr.[129]

[...] Les Valaques desquels descendent les Mégaloviachites, ou grands Valaques, paraissent être les dernières hordes des barbares, contre lesquelles les empereurs grecs eurent à soutenir des guerres, avant le dé bordement des Turcs dans l'Orient. Le nom de Bulgares, Όυννοδουνσουλγαροι Hunnobundobulgari, à l'époque dont Nicétas en parle dans ton histoire se confond avec celui de Valaques, dont il ne semble faire qu'une même nation, quoique très differente par le langage, la physionomie et les habitudes. C'est à ces derniers qu'il attribue les actes d'armes, les invasions et les ravages qui signalèrent l'ère malheureuse des faibles monarques de Constantinople, jusqu'à la fin tragique de Baudoin. Mais il ne dit rien, et nous n'avons que des données incertaines sur lorigine du peuple Valaque, qui se perd dans l'impénétrable obscurité des siècles. L'étymologie même de leur nom est un objet de doute parmi les savants. C'est en vain que le pape Pie U, plus connu dans les lettres sous le nom d'Aeneas Sylvius , croit que leur nom vient de Flaccus , général romain, qui subjugua les Moesiens et dispersa les Gètes; l'opinion du savant pontife ne reposant que sur le témoignage d'Ovide, qui n'a aucun trait direct avec la question, n'est ni prouvée, ni même vraisemblable.

---

[129] Pouqueville, F.-C.-H.-L., *Voyage de la Grèce,* Livre VI, Chapitre I, Ez Firmin Didot, Père et Fils, Paris, 1826, p.326-332

L'auteur anonyme de l'histoire de Moldavie assure que le nom de Valaqucs est le même que plusieurs nations donnent aux Italiens et aux Romains, desquels ils tirent leur origine. Les Allemands, par exemple, appellent également les uns et les autres Welsch, nom que Voltaire, dans ses traits ironiques , applique aux Français qui peuvent à bon droit se glorifier de celui qu'ils ont illustré. Les Polonais, de leur côté, donnent aux Italiens le nom de Wloch, et aux Valaques, celui de Woloch. Les Hongrois nomment enfin les Italiens Olach, et les Moldaves et Valaques Oulach, l'Italie, Wlochazeme, et la Valachie, Woloschazeme. Peyssonel[130], qui examine ces homonymies avec la sagesse caractéristique de ses recherches, penche, avec l'Anonyme de l'Histoire de Moldavie, pour l'opinion que les Valaques sont d'origine romaine. Ce suffrage d'un écrivain voyageur, instruit dans la langue de ces peuples, est d'un grand poids, et ce qui le confirme, c'est la langue des Valaques, qui, toute altérée et mélangée d'idiomes des peuples barbares, a cependant conservé le fonds, l'ordre, le rhythme et la syntaxe du latin. Quant à l'etymologie du nom de ce peuple, j'y donnerai la signification qu'il y attache lui-même, savoir, celle de Vlach, qui signifie *pasteur* ou *nomade*.

Les Mégaloviachites, qui habitent de nos jours les hautes montagnes du Pinde, que Nicétas appelle les Météores de la Thessalie, tels que ceux des cantons de Malacassis et d'Aspropotamos, se prétendent, sans fournir aucune preuve historique, descendants des débris de l'armee de Pompée, qui le réfugièrent dans les montagnes de Thessalie après la bataille de Pharsale. D'autres d'entre eux croient être la postérité d'une colonie sortie des Abruzzes; et ils donnent pour raison de cette tradition, que les Valaques Aspropotamites se

---

[130] Peyssonel, *Histoire des peuples barbares*

qualifient encore de Bruzzi-Vlachi[131]. Enfin la même opinion est commune aux Valaques Perrhébiens, qui habitent Mezzovo, une partie du canton de Zagori, de la Livadie, de l'Attique, et qu'on trouve jusqu'en Morée.

Les Valaques Massarets ou Dassarets, qui restaurèrent Moschopolis, à laquelle ils donnèrent le nom de Voschopolis, *ville des pasteurs*, à cause de leur titre de *Vlach* cette valeureuse peuplade, dont, les tribus sont disséminées dans les cantons de Caulonias, de Ghéortcha, et jusqu'au voisinage de Durazzo, sont, à les entendre, la postérité d'une colonie établie par Quintus Maximus, dans la Taulantie, ou Musaché, d'où ils seraient passés dans les monts Candaviens, au temps des invasions des barbares. Pour ce qui est des tribus Valaques Voisines du Parnasse et du Céphisse de la Phocide, elles prétendent avoir une origine commune avec les Mégaloviachites; et toutes en général revendiquent avec orgueil le nom de Romoûnis, ou Romains.

Je ne sais à quelle époque précise les Valaques se sont établis dans le Pinde, ni pourquoi les Grecs les ont surnommés Mégaloviachites. Cependant la première partie de ce problème se résoudrait, si on pouvait admettre en preuve leurs versions populaires; et ils y seraient à ce titre depuis une haute antiquité. Mais si on veut qu'ils soient venus des bords du Danube, ils sont modernes dans la Dolopie, car on ne parle guère même des Valaques comme nation, avant le dixième siècle. A cette époque de confusion, on les voit aux prises avec les empereurs grecs, incendiant et désolant les plus belles contré es de la Thrace et de la Macédoine. Parfois vaincus, et plus souvent vainqueurs, ils brillent par des traits de courage et de férocité. Unis aux Comains et aux Scythes,

---

[131] *Vlachs Bruzzi*, or, as Pouqueville says, *Valaques Brutiens*, coming from the slaves and shepherds of Lucanians, having rebelled against their masters.

ils descendent, comme des torrents dévastateurs, des sommets du mont Hemus et du Rhodope. Serrés, Philippopolis, Teruobe, Rodosto, éprouvent leurs fureurs; et l'Orient, épouvanté, tremble au seul bruit de leur nom. Ils fomentent toutes les révolutions pour y prendre part; et ils se mêlent aux convulsions sanglantes de l'état, afin de le démembrer et de s'en partager les lambeaux. Enfin, au mois de mars 1205, ils portent un coup fatal à ce fantôme d'empire que les Latins voulaient soutenir. Ils paraissent à la vue du camp français qui assiégeait Audrinople; ils le harcèlent, et attirent nos impétueux guerriers dans une embuscade où leur avant-garde, taillée en pieces, expie la faute de la valeur inconsidérée. Le comte de Blois, chef des braves, y perd la vie; et l'empereur Baudoin, fait prisonnier, est traîné à Ternobe devant le roi des Bulgares, qui lui fait subir la mort la plus affreuse. L'Europe chrétienne frémit à la nouvelle de cet horrible événement. La pitié du père commun des fidèles s'en émut; et ce fut seulement vingt-trois ans après ce grand désastre que Grégoire IX osa députer son légat, l'évéque de Strigonie, vers leur roi Borris. Cet envoyé de paix trouva les Bulgares, ou Valaques payens, habitant sous des tentes de feutre; et il eut la douleur de voir échouer sa mission évangélique; tant ces fiers courages, qui ne connaissaient que le fer et la guerre, étaient endurcis. Ils méprisèrent l'envoyé de paix. Ils l'insultérent lorsqu'il leur parla des saints et des peines de l'enfer, en lui répondant qu'*ils ne redoutaient ni les idoles, ni les esprits, et qu'ils n'adoraient que la force ei la valeur.*

Descriptio Europae Orientalis

1 . Notandum est hic quod inter Machedoniam, Achayam et Thesalonicam est quidam populus ualde magnus et spaciosus qui uocantur Blazi, qui et olim fuerunt Romanorum pastores, ac in Ungaria, ubi erant pascua Romanorum, propter nimiam terre uiriditatem et fertilitatem olim morabantur. Sed tandem ab Ungaris inde expulsi, ad partes illas fugierunt; babundat enim caseis optimis, lacte et carnibus super onmes nationes. Terram enim horum Blachorum que est magna et opulenta exercitus domini Karuli qui in partibus Grecie moratur fere totam occupauit et ideo conuertit se ad regnum Thesalonicense et actu mari terraque, expugnant ciuitatem Thesalonicensem dictam cum regione circumadiacente.

2 . Et est notandum, quod regnum Vngarie olim non dicebatur Vngaria, sed Messia et Panonia. Messia quidem dicebatur a messium prouentu[2]), habundat enim multum in messibus, Pannonia dicebatur etiam a panis habundantia; et ista consequenter se habent, ex habundantia enim messium

sequitur habundantia panis Panoni autem, qui inhabitabant tune Panoniam, oirmes erant pastores Romanorum, et habebant super se decern reges potentes in tota Messia et Panonia; deficiente autem imperio Romanorum egresi sunt Vngari de Sycia provinciá et regno magno, quod est ultra Meotidas paludes, et pugnauerunt in campo magno, quod est inter Sicambriam et Albam Regalem cum X regibus diclis et optinuerunt eos et in signum uictorie perpetuum erexerunt ibi lapidem inarmoreum permaximum, ubi est scripta prefata uictoria, quiad hue perseuerat usque in hodiemum diem

Trad. Popa-Lisseanu, *Izvoarele istoriei românilor*, Vol. II, Descrierea Europei Orientale de geograful Anonim, G., Tipografia Bucovina, Bucureşti, 1934, p.7

[...] Here are the various passages of Hungarian and foreign chroniclers who find out the existence of these shepherds in Pannonia. And we also quote foreign writers, because Hungarian historians claim that only in the Hungarian national springs these Roman shepherds are spoken of. [2]).

1. Anonymous notary:

a) *Rex Atliila... de terra scitliica descendens cum valida manu in terram Pannonie venit, et fugatis Romanis regnum obtinuit* (c. 1).

b) *Quam terram (Pannoniam) habitarent Sclavii, Bulgarii et Blachii ac pastores Romanorum. Quia post mortem Athile regis terram Pannonie Romani dicebant pascua esse, eoquod greges eorum in terra Pannonie pascebantur. Et iure terra Pannonie pascua Romanorum esse dicebatur, nam et modo Romani pascuntur de bonis Ungarie* (c. 9).

c) Dicebant enim eis sic, quod terra ilia (Pannonia) nimi bona esset, et ibi confluèrent nobilissimi fontes.... Et mortuo illo préoccupassent Romani principes terram Pannonie usque ad Danubium, ubi collocassent pastores suos (c. 11).

2. Odo of Deuil, French chronicler, tells us:

Terra hec (Ungaria) in tantum pabulosa est, ut dicantur in ea pabula Julii Caesaris extitisse[8]!.

3. Thomas of Spalato, in Dalmatia, states:

Haec regio dicitur antiquitus fuisse pascua Romanorum [4]).

4. Richardus, monk from the order of the praedicators, writes in his report:

Inventum fuit in Gestis Ungarorum Christianorum quod esset alia Ungaria maior... de qua septem duces... venerunt in terram que nune Ungaria dicitur, turn vero dicebatur pascua Romanorum, quam mhabitandam pre terris ceteris elegerunt, subiectis sibi populis, qui tune habitabant ibidem [5]).

5. Simon de Keza, in two passages of his chronicle, tells us:

a) *Pannonie, Pamfilie, Macedonie, Dalmacie et Frigie civitates que crebris spoliis et obsedionibus per Hunos erant fatigate, natali solo derelicto, in Apuliam per mare Adriaticum de Ethela licentia impetrata transierunt, Blackis, qui ipsorum fuere pastores et coloni remanentibus sponte iu Pannonia*[6]).

b) *Postquam aulem filii Ethele in prelio Crumhelt cum gente scilbica fere quasi deperissent, Pannonia extitit X annis sine rege, Sclavis tantum modo, Grecis, Teutonicis, Messianis et Ulahis advenis remanentibus, in eadem qui vivente Ethela populari servicio sibi serviebant*[7]).

These two passages of Simon de Keza's chronicle, with small variants, are also found in Chronicon Pictum Vindobonense[8]), in Chronicon Posoniense[9]), and in Chronicon Dubnicense[10]).

The data of these chronicles are now confirmed by the Anonymous Geographer from 1308 who did not know the Wallachians of Transylvania. As the appointment of Vlach has come, in time, an appellation, meaning 'shepherd' this is true, but only for a late epoch, after the twelfth century, and not for the first centuries of Hungarian history, not for the early times of the Arpadian dynasty.[11]).

That the Romanians at least during the time of Ladislau IV of Cumanus were not pastors, but a settled people, this results not only

288

from the fact that, in 1210, the Romanians took part in the expedition of Count Joacliim against Ascenus 'Burul' of Vidin, the capital of the Romanian-Bulgarian empire, associatis sibi Saxonibus, Olacis, Siculis et Bissenis *), and later, against the Czechs, in the Bela IV army, we innumerate the multitude of innumeram multitudinem Cumanorum, Ungarorum et diversorum Slavorum Siculorum quoque et Valachorum, but also from the mention that is made about them in the medieval songs that have been preserved until today. Thus the anonymous poet of the songs of the Nibelungi, by the twelfth century, Rudolf of Ems, from the middle of the thirteenth century in his Weltchronik, Jansen Enikel, from the thirteenth century, also in his work entitled Weltchronik[9] and then Stirian Ottocar, from the thirteenth century in the poem to the Österreichische Reimchronik, and others talk to us with enough details about the Vlachs who, at that time, could not be some nomadic shepherds.

1) Almus himself is sometimes called king, rex, although the chronicles designate him as the duke, *dux*.

2) Cf. los. Deer, *Ungarn in der Descriptio Europae Orientalis, in Mitteilungen des Österreich. Instit. f. Geschichtsforsch.* XLV, 1931.

3) Odo of Deogilo — (Deuil, near Paris) — *Liber de via sancti sepulchri p. 62.* Schunemann believes to be a confusion between *Cereris* and *Caesaris, Ung. Jahrbuch.* 1926.

4) Thomas Spalatensis, *Historia Salonitana*, at Schwandtner, *Scrip. rer. Hung.* III. p. 549.

5) Richardus, De facto Ungarie magne, at Endlicher, Rerum Hungaricarum Monumenta Arpadiana, p. 248.

and p. 70: Zaculi Hunorum sunt residui qui... cum Blackis in montibus confinii

6) Florianus, Magistri Simonis of Keza, *Gesta Hungarorum II* p. 65; cf. sortem habuerunt; unde Blackis commixti literis ipsorum uti perhibentur.

7) Florianue, *ibidem*, p. 70.

8) Florianus, *ibidem*, p. 114 şi 120.

9) Florianus, *Chronica minora* IV, p. 15 şi 21.

10) Florianus, Chronica Dubnicense III, p. 17 şi 23.

11) Cf. Silviu Dragomir, *Vlachii şi Morlacii* p. 51 seq.

Ibid., p.10-11

The Romanians of the Balkan Peninsula

by Mihai Eminescu

[...]

There is no state in Eastern Europe, there is no country from the Adriatic to the Black Sea that does not include pieces of our nationality. Starting with the shepherds in Istria, from the morlacs in Bosnia and Herzegovina, we find step by step the fragments of this great ethnicity in the mountains of Albania, Macedonia and Thessaly, Pindos as well as in the Balkans, Serbia, Bulgaria, Greece, till the walls of Athens, then beyond Tisza, from all over the Dacia Trajan to Dniester, near Odessa and Kiev.

While the Russians have the greatest care for the most insignificant tribes of the great Slav family, while the Germans are persisting through their consular authorities for their most insignificant colonies in the Orient, and each Western nation develops a special care for Its own nationalities in these places, we alone struggle in internal fights for the best possible form of human organization, lacking an ideal of culture, but at the most political ideals that do not match our powers and which, instead of giving birth to the deeds, will at most be the cause of dangerous adventures.

On the occasion of the Congress in Berlin, almost all peoples of the Balkan Peninsula gave a sign of life, only

Transdanubian Romanians did not. The cause is easy to understand. All other population fragments are linked to culture with those political centers created by their nationalities. Greeks in European Turkey read and write the language spoken in Athens; Serbs in Turkey are well aware of the institutions and culture of their free counterparts; only us, with our way of seeing, we are strangers in the Orient, and we remain misunderstood even for those of the same family. How could we in other ways explain the really strange phenomenon that such important parts of the ancient population as the Romanians in Thessaly and Macedonia are not to give any sign of life, with all their brilliant past, despite the fact that they preserved and defended their tongue and doctrine better than the Slavs, many of whom were now Greeks, than the Albanians, of whom many also became Turks.

'The Vlachs of Thessaly, says Flammerayer in his Fragments of the East, they call themselves Romanians, as their countrymen in the Danubian Principalities, speak of a broken Italian, and live in the brains of the Pindus mountains and on both sides of it, in the peoples from which Peneios originates and the Affluent rivers, where the Byzantine history of the eleventh century first remembers them. Whether remains of the Roman military colonies or Latinized native barbarians, they stretch and branched along the mountain range through Upper Macedonia up to the Balkans and once they were in touch with their border guards on the left bank of the Danube. They guard and dominate the gates between Thessaly and Albania, and Mezzovo, a city built of stone just in the brain of the mountains, where from one side and the other the passage descends in opposite directions, is the head of the Tesalian Romanians.

Malakasi, Lesinita, but especially Kalarites, Kalaki and Klinovo, and twenty-one villages in the Pind's pontoons, and besides them, they are also of this people, who due to the harsh temperature of their homeland, deals with agriculture, deals with agriculture, but more so with the culture of cattle and the witches, this in a great and successful style, that through the richness of their flocks of sheep they are all famous in Rumelia.

In the winter, when the oats cover the mountain heights, they move their flocks in the valleys with a milder climate and graze them, nomadizing on the grassy grasslands, even within the free Greece, and when the spring returns, the black tortuous villages of the wandering Romanian shepherds disappear from the plain, for they return to the mountain.

Sobral, having the instinct of marriage and industry, Romanians are looking at these qualities far superior to those who speak Greek; but they are inferior to the Greco-Slavs in spirit and in cunningness. However, these shepherds, simple and straightforward, have an eminent ability to work in metal. Guns and armorworks worked in gold and silver we admire at the Arnauti and Palicari came out of the works of the Vlachs. Likewise, raincoat hooded jackets and well-known in all Mediterranean port towns like Cappa, Greco and Marinero are mostly a product of the Vlach drapers industry. Vlachs Grocers and guilds are found in all European European cities, and even in Hungary and Austria, where their love of profit lead them. That they are also good at the merchandise in the sea proves the rich Sina of Vienna, a Vlach born, if we do not deceive, in Klinovo, or yet in one of the Pindos settlements named above.

From this traveling life, the general familiarization of the Vlachs with the Neo-Greek dialect, the dialect that they use also in the church, which forms the common means of understanding and linking the different nationalities from the two sides of the Aegean Sea. Women in many villages are only able to understand Romanian. Like all the mountain inhabitants, the Vlach cannot forget the homeland even in the farthest countries, and even returns when old in Pindos with what he has gathered through the labor of a whole life to be buried in the same land as his ancestors rest..

But the people of the Vlachs, so peaceful today and dedicated only to work and gain, have not always been enlivened by such a quiet spirit, nor have they been confined and confined to its present settlements through the western mountains of Thessaly. The tessalian Vlachs, as their Albanian neighbors later, have had their brilliance and political age, short and passable as the size of the Tebanians; but in the Byzantine era, not without meaning. Next to the Vlacho-Libadi and Vlacho-Iani villages, which still exist today in the Southern promontories of the Cambunici Mountains, not far from Târnova, Ana Comnenus (1083) mentions a Vlach trade town, Exebas, in the valleys of Mount Pelion at the eastern edge of Thessaly; and Benjamin of Tudela, who in the twelfth century traveled through Greece, says that in the South Zitum was the city of the border and the entrance to the country of the Vlachs. Like the Peloponnesus, Thessaly had lost its ancient name in the Middle Ages, and it was called, for hundreds of years afterwards, only Megaly Blahia, proper Greater Valahia, to distinguish it from Acarnania and Etolia, which after the Byzantine George Phrantzes was called Little Valahia. George Pachymeres, the court historian of the first Palaeologist Mihail, makes it clear that the tesalians, once commanded by Achiles and formerly called Elinas, are called in his days of great Vlachites (G. Pachymeres în Mich. Paleol(og), 1,30).

Nicetas of Chonae borders Megale Vlachia on the mountain ring and the hill country that rises above the plains, and the central plains, inhabited by the cowards and the non-warriors Greek-Slavs, like to call it Thessaly. But does not the

Rabbi Benjamin make clear how the Vlachs live in the mountains and descend to the Greek region to pillage them? In sprinter, that traveler compares them with deer, their warrior courage is unbearable, and no king has been able to resist them.

The man of Tudela was well aware of the impressions of his age, for soon after the journey of Rabbi Benjamin (1186), all Romanians in the Pindus mountain chain, up in the valleys of the Balkans, rose under their rulers Peter and Asan against the oppressive and bloody reign of the thieves of the Byzantine Court, founded a kingdom with the capital Târnova on the Northern Emul (Balean). The most Southern edge of the Romanian-Bulgarian kingdom were the mountains of Tesalia, under a independent captain called (Megas) Vlachos (that is, the Great Romanian) and shining under that name in the contiguous chronicles of the Franks and Byzantines.'

That's what Fallmerayer says:

We also know that all these Romanians had taken Thrace, Macedonia and Thessaly, that they overcame the Greek armies and those of the Latin kingdom of the Orient, that they captured Baldovin I, killed the flower of the Western knights, that the Asanids were recognized by the pope as the royal dynasty of Europe, as the legitimate kings of Blacorum et Bulgarorum, in a word that this fragment of people, so ignored today, when neither in the journalism nor in the congress had been mentioned, has a brilliant past from his own bravery over the enemies with far superior in culture and the art of war..

And yet these people, in our country, even among their fellow countrymen, were known only under the ridiculous nickname of kutovlahi. And while the possessions of our boyars and monasteries worshiped without grieving at Greek churches and were used for Greek purposes, there was not one whole church in that region at least when the Romanian language was heard. [...]

Ziarul 'Timpul,' III, nr. 211, 26 Septembrie 1878, p. 1-2

Instructions given by Titu Maiorescu

to the Ambassador to London on the meeting of the
Great Powers' Ambassadors, London, December
1912ambasadorului la Londra referitoare la
reuniunea ambasadorilor Marilor Puteri, Londra,
decembrie 1912

Minister of Foreign Affairs to the Plenipotentiary Minister of Romania in London

Bucharest, Saturday, 15/28 December 1912

Minister,

With regard to the issues that will be the subject of your mission in London, I would like to draw your attention to the following two essential points, giving you the views of the Romanian Government, to guide you in your negotiations.

I. - Reunion of the Great Powers' Ambassadors in London:Romania did not ask to attend the meeting, but - of course - would be pleased to be offered and to receive this, on a straightforward basis, on an equal footing.

Regarding the attitude of the Powers in this matter, I must tell you that Germany and Austria-Hungary have asked for Romania's participation. Italy supported the request.

With a possible participation of Romania at the Ambassadors meeting, you will seek to defend, above all, the interests of the Aromanians. In this regard, it may be an autonomous Macedonia and an Albania, possibly an Albania as large as possible. In your negotiations, you will insist that the Balkan states, and especially Greece, respect the schools and churches of the Aromanians, and not embarrassing at all the establishment of their Episcopate.

And the interests of the Romanian monks at Mount Atos can form the subject of your observations, guiding you through the attached special special note and putting you, in particular for this matter, in closer contact with the Russian Ambassador.

The Danube-Adriatic railway line may also be important to us, which in the examination will also take into account the proposal - as outlined below - what could be done for Bulgaria to throw a bridge over the Danube.

It may be advisable that, if your negotiations with Mr. Daneff do not reach a satisfactory result, you may also reach in the Ambassadors meeting our discussion with Bulgaria, and possibly the danger of an unexpected recalcitrance of the Bulgarian Government.

II. - Negotiations with Bulgaria, especially with Mr. Daneff:

You know the circumstances in which Mr. Daneff's mission was carried out in Bucharest.

Let me briefly tell you the conversations he had with the Romanian Government:

1 ° As for the Macedonian Romanians, Mr. Daneff assured us that Bulgaria will respect their schools and churches in the occupied localities and admits the establishment of an Episcopate for Macedonian Aromanians. No objection was raised to the direct

subsidization by the Romanian State of the Aromanian culture institutions.

2° Dobrogea border:

Given the current friendship relations between Bulgaria and Romania, which we want to be even closer in the future, there can be neither territorial compensations nor strategic lines, but only a border correction, which would presents a guarantee and a certainty for the future about the sincerity of mutual friendship, replacing - at the moment when such the Berlin tract is touched by a great deal - the current boundary imposed by that tract, against our will, and removing any misunderstanding between the two States.

For this purpose, Romania must reach a border line which, starting from the Danube and reaching the Black Sea, should be drawn as far South as possible from the current border. There are four major routes on four maps that you have at your fingertips.

The first route - the strategic line - you can portray it as representing the desires of our military circles, but it should be removed from the discussion for the reasons already set.

You will insist on the others. From the telegram that I will send these days, you will see that we could not admit the route which, going a little westwards from Silistria, would reach Balcik, leaving Dobrich on the side of Bulgaria.

Apart from the benefits that the Balkan states, and Bulgaria in particular, have drawn from Romania's neutrality, we are ready to grant other benefits, so that the Bulgarian Government can justify before Romania the border correction that is due to Romania.

The question that we put up today was 'sous-entendues' when we kept the neutrality that allowed, or in any case helped, the success of the Bulgarians towards the Turks.

And now, Romania will be more willing to exert its influence on the Powers in the sense of Bulgaria's desires, the more she will know her better to protect the interests of the Aromanians from *her mastery and to offer us sincere and spontaneous friendship Border correction.*

A bridge over the Danube could finally give satisfaction to an older aspirations of Bulgaria, being in the interests of its economic interests.

These are the points on which I wanted to draw your attention.

I will not miss giving you detailed instructions on each point as soon as they are needed after your negotiations.

T. MAIORESCU

(MINISTERUL AFACERILOR STRĂINE. Documente diplomatice. Evenimentele din Peninsula Balcanică. Acțiunea României. 20 Sept. 1912—1 Aug. 1913, București, Imprimeria Statului, 1913, P. 14-15)

Pro-memory submitted to the Foreign-Office

on the request of Sir Edward Grey.

Londres, le 14/27 Mars 1913.

Le Gouvernement Roumain voit avec satisfaction la création d'un État Albanais indépendant, d'autant plus qu'il espère que, par les mesures que les Grandes Puissances garantes voudront bien prendre, l'individualité des nombreux Roumains qui y seront incorporés, sera sauvegardée. Dans ce but, les limites de la future Albanie devraient être tracées de telle manière que non seulement cet État soit à l'abri de toutes difficultés ultérieures avec ses voisins, mais qu'aussi la population roumaine, qui est la plus compacte au Sud-Est de l'Albanie, soit conservée intacte dans les confins de l'État Albanais.

La région comprise entre les villes Janina, Metzovo, Grebene et le mont Gramos, est habitée par une population en majorité roumaine, qui peut être évaluée à plus de 80.000 habitants et qui est groupée dans 36 villages et bourgades, dont les principales sont: Samarina, Avdela, Perivoli, Crania, Labanitza, Seracu, Perivoli, Laïsta, Leshnitza, Breaza, Turia, Medjidie, etc.

Il y a lieu de faire remarquer que les deux versants du Pinde, depuis le mont Gramos jusqu'à l'Agrafa, sont occupés en majorité par les Roumains.

Une partie de cette population a été annexée à la Grèce après le Traité de Berlin. Les Roumains ont protesté alors contre cette annexion. Ce serait injuste de permettre de nouveau la séparation du tronc compacte des Roumains et de les annexer à d'autres États que l'Albanie. La Roumanie est d'avis que leur individualité nationale sera mieux conservée dans un État albanais indépendant, sous la garantie et le contrôle des Grandes Puissances, dont les limites devraient être fixées d'une manière aussi indiscutable que possible, pour éviter les troubles à l'avenir dans ces contrées.

Le Gouvernement Roumain considère que les meilleures frontières naturelles pour l'Albanie du Sud seraient les montagnes du Zagori (Mitchikeli et Papingo) ; la vallée de la rivière Inahos jusqu'à son confluent avec la rivière Arta (Arachtos), jusqu'à la source de cette rivière à Joug (Zygos) ; d'ici à Metzovo et, en suivant la frontière actuelle de la Grèce, jusqu'à la rivière Venetico, et de là jusqu'à son confluent avec la rivière Bistritza (Aliacmon-Indje-Sou), suivre le cours de la Bistritza vers Darda, Gramoste, Koritza, jusqu'au lac de Prespa.

La population de ces confins est en grande partie roumaine, musulmane (Roumains mahométans-Vlachades), albanaise et, une minorité, grecque.

Pour la sauvegarde de l'individualité nationale des Roumains de ces contrées, qui seront incorporés à l'Albanie, les Grandes Puissances voudront bien inscrire, non seulement dans le Traité international qui remplacera le Traité de Berlin, mais aussi dans la Constitution ou statut organique de l'Albanie, le principe que dans l'administration de toutes les localités où la majorité serait roumaine, de même que dans toutes les églises et écoles roumaines, la langue usuelle soit roumaine.

Le nouvel État albanais devrait garantir une autonomie administrative et communale et, autant que possible, politique aux Roumains de l'Albanie, en ne mettant aucun obstacle au fonctionnement du chef religieux roumain des cantons habités par des Roumains.

L'État roumain pourra comme par le passé, subventionner les institutions de culture roumaines de l'Albanie, sans aucune restriction de la part de l'État albanais.

N. Mișu

(*Ibid.*, P. 91-92)

ROUMANIE, GRECE, MONTENEGRO, SERBIE, BULGARIE.

Traité de paix signé à Bucarest

le 28 juillet / 10 août 1913, suivi de deux Procès – verbaux d'échange des ratifications. Publication officielle. Bucarest 1913.

TRAITE DE PAIX

Leurs Majestés le Roi de Roumanie, le Roi des Hellènes, le Roi de Monténégro et le Roi de Serbie, d'une part, et Sa Majesté le Roi des Bulgares, d'autre part, animés du désir de mettre fin à l'état de guerre actuellement existant entre Leurs pays respectifs, voulant, dans une pensée d'ordre, établir la paix entre Leurs peuples si longtemps éprouvés, ont résolu de conclure un Traité définitif de paix. Leurs dites Majestés ont, en conséquence, nommé pour Leurs Plénipotentiaires, savoir:

Sa Majesté le Roi de Roumanie:

Son Excellence Monsieur Titus Maïoresco, Son Président du Conseil des Ministres, Ministre des Affaires Etrangères; Son Excellence Monsieur Alexandre Marghiloman, Son Ministre des Finances; Son Excellence Monsieur Take Ionesco, Son Ministre de l'Intérieur; Son Excellence Monsieur Constantin G. Dissesco, Son Ministre des Cultes et de l'Instruction Publique; Le Général de division aide de camp C. Coanda, Inspecteur général de l'artillerie, et Le Colonel C. Christesco, Sous-chef du grand état-major de Son armée.

Sa Majesté le Roi des Hellènes:

Son Excellence Monsieur Eleftéris Veniselos, Son Président du Conseil des Ministres, Ministre de la Guerre; Son Excellence Monsieur Démètre Panas, Ministre Plénipotentiaire; Monsieur Nicolas Politis, Professeur de droit international à l'Université de Paris; Le Capitaine Ath. Exadactylos, et Le Capitaine C. Pali.

Sa Majesté le Roi de Monténégro:

Son Excellence le Général Serdar Yanko Voukotitch, Son Président du Conseil des Ministres, Ministre de la Guerre, et Monsieur Jean Matanovitch, Ancien Chargé d' Affaires de Monténégro à Constantinople.

Sa Majesté le Roi de Serbie:

Son Excellence Monsieur Nicolas P. Pachitch, Son Président du Conseil des Ministres, Ministre des Affaires Etrangères; Son Excellence Monsieur Mihaïlo G. Ristitch, Son Envoyé Extraordinaire et Ministre Plénipotentiaire à Bucarest; Son Excellence Monsieur le Docteur Miroslaw Spalaïkovitch, Envoyé Extraordinaire et Ministre Plénipotentiaire; Le Colonel K. Smilianitch, et Le Lieutenant Colonel D. Kalafatovitch.

Sa Majesté le Roi des Bulgares:

Son Excellence Monsieur Dimitri Tontcheff, Son Ministre des Finances; Le Général-Major Ivan Fitcheff, Chef de l'état-major de Son armée; Monsieur Sawa Ivantchoff, docteur en droit, ancien Vice-Président du Sobranié; Monsieur Siméon Radeff, et Le Lieutenant Colonel d'état-major Constantin Stancioff.

Lesquels, suivant la proposition du Governement Royal de Roumanie, se sont rénuis en Conférence à Bucarest, munis de pleins pouvoirs, qui ont été trouvés en bonne et due forme.

L'accord s'étant heureusement établi entre eux, ils sont convenus des stipulations suivantes:

ARTICLE PREMIER

Il y aura, à dater du jour de l'échange des ratifications du présent Traité, paix et amitié entre Sa Majesté le Roi de Roumanie, Sa Majesté le Roi des Hellènes, Sa Majesté le Roi de Monténégro, Sa Majesté le Roi de Serbie et Sa Majesté le Roi des Bulgares, ainsi qu'entre Leurs héritiers et successeurs, Leurs Etats et sujets respectifs.

ARTICLE II

Entre le Royaume de Bulgarie et le Royaume de Roumanie, l'ancienne frontière entre le Danube et la Mer Noire est, conformément au procès-verbal arrêté par les Délégués militaires respectifs et annexé au Protocole No 5 du 22 juillet (4 août) 1913 de la Conférence de Bucarest, rectifiée de la manière suivante:

La nouvelle frontière partira du Danube, en amont de Turtukaïa, pour aboutir à la Mer Noire au Sud d'Ekrene.

Entre ces deux points extrêmes, la ligne frontière suivra le tracé indiqué sur les cartes 1/100.000 et 1/200.000 de l'état-major roumain, et selon la description annexées au présent article.

Il est formellement entendu que la Bulgarie démantélera, au plus tard dans un délai de deux années, les ouvrages de fortifications existants et n'en construira pas d'autres à Roustchouk, à Schoumla, dans le pays intermédiaire, et dans une zône de vingt kilomètres autour de Baltchik.

Une commission mixte, composée de représentants des deux Hautes Parties contractantes, en nombre égal des deux côtés, sera chargée, dans les quinze jours qui suivront la signature du présent Traité, d'exécuter sur le terrain le tracé de la nouvelle frontière, conformément aux stipulations précédentes. Cette commission présidera au partage des biens-fonds et capitaux qui ont pu jusqu'ici appartenir en commun à des districts, des communes, ou des communautés d'habitants séparés par la nouvelle frontière. En cas de désaccord sur le tracé et les mesures d'exécution, les deux Hautes Parties contractantes s'engagent à s'adresser à un Gouvernement tiers ami pour le prier de désigner un arbitre dont la décision sur les points en litige sera considérée comme définitive.

ARTICLE III

Entre le Royaume de Bulgarie et le Royaume de Serbie, la frontière suivra, conformément au procès-verbal arrêté par les Délégués militaires respectifs et annexé au Protocole No 9 du 25 juillet (7 août) 1913 de la Conférence de Bucarest, le tracé suivant:

La ligne frontière partira de l'ancienne frontière du sommet Patarica, suivra l'ancienne frontière turco-bulgare et la ligne de partage des eaux entre le Vardar et la Strouma avec l'exception que la haute vallée de la Stroumitza restera sur territoire serbe; elle aboutira à la montagne Belasica, où elle se reliera à la frontière bulgaro-grecque. Une description détaillée de cette frontière et son tracé sur la carte 1/200.000 de l'état-major autrichien, sont annexés au présent article.

Une commission mixte, composée de représentants des deux Hautes Parties contractantes, en nombre égal des deux côtés sera chargée, dans les quinze jours qui suivront la signature du présent Traité, d'exécuter sur le terrain le tracé de la nouvelle frontière, conformément aux stipulations précédentes.

Cette commission présidera au partage des biens-fonds et capitaux qui ont pu jusqu'ici appartenir en commun à des districts, des communes, ou des communautés d'habitants séparés par la nouvelle frontière. En cas de désaccord sur le tracé et les mesures d'exécution, les deux Hautes Parties contractantes s'engagent à s'adresser à un Gouvernement tiers ami pour le prier de désigner un arbitre dont la décision sur les points en litige sera considérée comme définitive.

ARTICLE IV

Les questions relatives à l'ancienne frontière serbo-bulgare seront réglées suivant l'entente intervenue entre les deux Hautes Parties contractantes, constatée dans le Protocole annexé au présent article.

ARTICLE V

Entre le Royaume de Grèce et le Royaume de Bulgarie, la frontière suivra, conformément au procès-verbal arrêté par les Délégués militaires respectifs et annexé au Protocole No 9 du 25 juillet (7 août) 1913 de la Conférence de Bucarest, le tracé suivant:

La ligne frontière partira de la nouvelle frontière bulgaro-serbe sur la crête de Belasica planina, pour aboutir à l'embouchure de la Mesta à la Mer Egée.

Entre ces deux points extrêmes, la ligne frontière suivra le tracé indiqué sur la carte 1/200.000 de l'état-major autrichien et selon la description annexées au présent article.

Une commission mixte, composée de représentants des deux Hautes Parties contractantes, en nombre égal des deux côtés, sera chargée, dans les quinze jours qui suivront la signature du présent Traité, d'exécuter sur le terrain le tracé de la frontière conformément aux stipulations précédentes.

Cette commission présidera au partage des biens-fonds et capitaux qui ont pu jusqu'ici appartenir en commun à des districts, des communes, ou des communautés d'habitants séparés par la nouvelle frontière. En cas de désaccord sur le tracé et les mesures d'exécution, les deux Hautes Parties contractantes s'engagent à s'adresser à un Gouvernment tiers ami pour le prier de désigner un arbitre dont la décision sur les points en litige sera considérée comme définitive.

Il est formellement entendu que la Bulgarie se désiste, dès maintenant, de toute prétention sur l'île de Crète.

ARTICLE VI

Les Quartiers généraux des armées respectives seront aussitôt informés de la signature du présent Traité. Le Gouvernement bulgare s'engage à ramener son armée, dès le lendemain de cette signification, sur le pied de paix. Il dirigera les troupes sur leurs garnisons où l'on procédera, dans le plus bref délai, au renvoi des diverses réserves dans leurs foyers.

Les troupes dont la garnison se trouve située dans la zône d'occupation de l'armée de l'une des Hautes Parties contractantes, seront dirigées sur un autre point de l'ancien territoire bulgare et ne pourront gagner leurs garnisons habituelles qu' après évacuation de la zône d'occupation sus-visée.

ARTICLE VII

L'évacuation du territoire bulgare, tant ancien que nouveau, commencera aussitôt après la démobilisation de l'armée bulgare, et sera achevée au plus tard dans la quinzaine.

Durant ce délai, pour l'armée d'occupation roumaine, la zône de démarcation sera indiquée par la ligne Sistov-Lovcea-Turski-Izvor-Glozene-Zlatitza-Mirkovo-Araba-Konak-Orchania-Mezdra- Vratza-Berkovitza-Lom-Danube.

ARTICLE VIII

Durant l'occupation des territoires bulgares les différentes armées conserveront le droit de réquisition, moyennant paiement en espèces.

Elles y auront le libre usage des lignes de chemin de fer pour les transports de troupes et les approvisionnements de toute nature, sans qu' il y ait lieu à indemnité au profit de l'autorité locale.

Les malades et les blessés y seront sous la sauvegarde des dites armées.

ARTICLE IX

Aussitôt que possible après l'échange des ratifications du présent Traité, tous les prisonniers de guerre seront réciproquement rendus.

Les Gouvernements des Hautes Parties contractantes désigneront chacun des Commissaires spéciaux chargés de recevoir les prisonniers.

Tous les prisonniers aux mains d'un des Gouvernements seront livrés au commissaire du Gouvernement auquel ils appartiennent ou à son représentant dûment autorisé, à l'endroit qui sera fixé par les parties intéressées.

Les Gouvernements des Hautes Parties contractantes présenteront respectivement l'un à l'autre, et aussitôt que possible après la remise de tous les prisonniers, un état des dépenses directes supportées par lui pour le soin et l'entretien des prisonniers, depuis la date de la capture ou de la reddition jusqu'à celle de la mort ou de la remise. Compensation sera faite entre les sommes dues par la Bulgarie à l' une des autres Hautes Parties contractantes et celles dues, et la différence sera payée au Gouvernement créancier aussitôt que possible après l'échange des états de dépenses sus-visés.

ARTICLE X

Le présent Traité sera ratifié et les ratifications en seront échangées à Bucarest dans le délai de quinze jours ou plus tôt si faire se peut.

En foi de quoi, les Plénipotentiaires respectifs l'ont signé et y ont apposé leurs sceaux.

Fait à Bucarest le vingt huitième jour du mois de juillet (dixième jour du mois d'août) de l'an mil neuf cent treize.

Signés:
Pour la Roumanie: (L.S.) T. Maïoresco, Al. Marghiloman, Take Ionesco, G. Dissesco, Général aide de camp Coanda, Colonel C. Christesco Pour la Bulgarie: (L.S.) D. Tontcheff, Général Fitcheff, Dr. S. Ivantchoff S. Radeff, Lt Colonel Stancioff Pour la Grèce: (L.S.) E.K. Veniselos, D. Panas, N. Politis, Capitaine A. Exadactylos, Capitaine C. Pali Pour le Monténégro: (L.S.) Général Serdar I. Voukotitch, Y. Matanovitch Pour la Serbie: (L.S.) Nik. P. Pachitch, M. G. Ristitch, M. Spalaïkovitch, Colonel K. Smilianitch, Lt Colonel D. Kalafatovitch

The description of Nikolaus Dumba

Of course, there are still hundred of us who have met him in the bloom of his manhood. He was a very handsome man (wunderschön), of course, one of the most beautiful of them that has ever lived. And this bodily quality, which appears so often intertwined with unfriendly spiritual properties and — as a result — It has a rather repellent, rather than attractive, influence - it has no repulsive taste. He was completely free of idiotic vanity, and in his relations with the people he was so natural and Viennese, so gemütlich Viennese (with goodness), and in the corner of his mouth he had such a fine roguish-Viennese smile, when he was conversing, that you forgot all that this man would be, and the beautiful man and the rich, and the one who occupies countless honorable posts, [...] and you only saw one, Dumba, the Viennese Viennese from Macedonia.

Something exotic is growing around him, though only around his exterior. The dark color of his face, the shining matte like the ebony, the dark night of his hair and his beard, the dark fire of his eyes-all without saying that the sunshine of that man had made him a more Southern sun. Inside, a Viennese, the best in the world, remained outside ... the luxurious copy of a man in the Balkans, the beautiful man in the Haemos mountains.

If Greek blood flowed in his veins, who can know? To know you in that maze of peoples of that peninsula, it passes almost over human powers. I gladly enjoyed the idea of a descendant of the remnant of the interesting and ancient people of the Aromanians or of the Macedonian-Wallachians, whom I have mentioned above, and of course, the Sina family.

TRANSSILVANICA, Biblioteca digitală, Biblioteca Centrală Universitară 'Lucian Blaga' Cluj-Napoca,
http://documente.bcucluj.ro/web/bibdigit/periodice/transilvania/1925/
BCUCLUJ_FP_279996_1925_056_005_006.pdf, p.12-3

BIBLIOGRAPHY

BOOKS

., Fontes Historiae Dacoromaniae, vol. I-IV, București, 1964-1982

A BUDAI EGYETEMI NYOMDA ROMÁN KIADVÁNYAINAK DOKUMENTUMAI 1780-1848. Budapest, 1982

ANDERSON, F. M., HERSHEY, A. S., Handbook for the Diplomatic History of Europe, Asia and, Africa 1870-1914, Washington, Government Printing Office, 1918

ARGINTEANU, IOAN D., Raportul despre mersul Liceului în curs de 25 de ani, Revista Lumina, Bitolia (Monastir), oct. 1905

ARION, V., PÂRVAN, V., PAPAHAGI, PERICLE, BOGDAN DUICA, G. - Românii și popoarele balcanice, București, 1913.

ARMBRUSTER, A., Romanitatea românilor. Istoria unei idei, București, 1972 (ediția franceză îmbunătățită, București, 1977).

BAKER, JAMES, La Turquie, trad. de J. de Caters, Paris, 1883

BERÉNYI, MARIA - Viaţa şi activitatea lui Emanuil Gojdu 1802-1870,
Societatea Culturală a Românilor din Budapest, Giula, 2002

BOIAGI, MIHAIL G., Gramatica românească sau macedo-română,
Bucureşti, 1915.

BOSCH, E. ET AL, Paternal and maternal lineages in the Balkans show
a homogeneous landscape over linguistic barriers, except for the
isolated Aromuns. Annals of Human Genetics, 2006, 70: 459–
487.

BOUÉ, AMI, La Turquie d'Europe, Paris, 1840.

BRAILSFORD, HENRY N., Macedonia: its Races and their Future,
London, 1903,

BRATIANU, I. I. C., România şi Peninsula Balcanică, Bucureşti, 1913.

CÂNDROVEANU, HRISTU, Aromânii ieri şi azi. Scrisul românesc,
Craiova, 1995.

CANTEMIR TRAIAN, Noi date istorice referitoare la istro-români,
Bucureşti, 1968.

CAPIDAN, TH., Macedoromânii. Etnografie, istorie, limbă. Bucureşti,
Fundaţia Regală pentru Literatură şi Artă, 1942.

COLSON F., Nationalité et régénération des paysans moldo-valaques,
E. Dentu, Paris, 1862

COMAS, D. ET AL, Alu insertion polymorphisms in the Balkans and the
origins of the Aromuns, Annals of Human Genetics, Volume 68,
Issue 2, , March 2004, p.120–127

DENSUŞIANU, OVIDIU, Viaţa păstorească în poezia noastră populară,
Bucureşti, 1922

DIAMANDI-AMINCEANU, VASILE, Românii din peninsula balcanică, Inst. De arte grafice 'TIPARUL UNIVERSITAR', București, 1938

DIONISIUS LASIC, O.F.M., Fr. Bartholomaei de Alverna, Vicarii Bosniae 1367-1407, quaedam scripta hucusque inedita, 'Archivum Francescanum Historicum', LV, 1962, 1-2,

DJUVARA, NEAGU, COORDONATOR, Aromânii - Istorie - Limbă. Destin. București, Fundația Culturală Română, 1996. Ediția a doua, București, HUMANITAS, nedatată.

DJUVARA, NEAGU, O scurtă istorie a românilor povestită celor tineri, Ed. Humanitas, București, 2010,

DR. BOTIȘ, TEODOR, Monografia familiei Mocioni, FUNDAȚIA PENTRU LITERATURĂ ȘI ARTĂ 'REGELE CAROL II', București, 1939

EMINESCU, MIHAI, articole pblicate în ziarul 'Timpul'

FOAIE PENTRU MINTE, INIMĂ ȘI LITERATURĂ, NR.18, 1861

GARRETT HELLENTHAL, GEORGE B. J. BUSBY, GAVIN BAND, JAMES F. WILSON, CRISTIAN CAPELLI, DANIEL FALUSH, AND SIMON MYERS, A Genetic Atlas of Human Admixture History, Science 14 February 2014: 343 (6172), 747-751. [DOI:10.1126/science.1243518]

HÂCIU, A., Aromânii. Comerț. Industrie. Artă. Expansiune. Civilizație, Tip. Cartea Putnei, Focșani, 1936.

IONESCU, NAE, Neliniștea metafizică, Editura Fundației Culturale Române, 1993

IONESCU-SACHELARIE, D., Despre viața păstorească și agricolă în trecutul nostru, București, 1941

IORGA, NICOLAE, CE ÎNSEAMNĂ POPOARE BALCANICE. Conferință ținută la Ateneul Romîn în ziua de 13 Decembre 1915, Neamul Românesc, Vălenii-de-Munte, 1916,

IORGA, NICOLAE, Istoria poporului românesc, trad. din lb. germ. de O.E. Ionescu, vol. IV, partea a II-a, Ed. Casei Şcoalelor, 1928

LAMBRU, SPIRU, Istoria tis Ellados, Atena, 1898

LAROUSSE, 'Istoria lumii de la origini până în anul 2000', ed. Olimp, București, 2000

LE COURRIER DES BALKANS, NR. 39, 1905

LUCII, IOANNIS, De regno Dalmatiae et Croatiae, Libri sex, Amsterdam, 1666

MAIDEN, MARTIN, 'Perfect pedigree. The ancestry of the Aromanian conditional', in Ashdowne, Richard and Finbow, Thomas (eds.), Oxford Working Papers in Linguistics, Philology and Phonetics, 9: 83-98, 2004.

MAIORESCU, TITU, România, războaiele balcanice şi Quadrilaterul, ed. a 2-a, București, Ed. Machiavelli, 1995.

MANTSU, YIANI, Juridical and political aspects regarding the minority of the Makedon-Aromâns in Albania and how they enjoy all the rights granted by the European and international norms, Conference on Minorities, Tirana, 13-14 Feb. 2013

MATZOTA, Eugene, Istoria aromânilor în date, Timișoara, 2011

MATZOTA, EUGENE, Românii contra aromânilor?, Brașov, 2016

MCNEILL, WILLIAM H., Mythistory and Other Essays. Chicago and London: The University of Chicago Press, 1986.

MCNEILL, WILLIAM H., 'Mythistory, or Truth, Myth, History, and Historians.' Presidential Address. The American Historical Review, 91, no. 1: 1-10, 1986.

MINISTERUL AFACERILOR STRĂINE. Documente diplomatice. Evenimentele din Peninsula Balcanică. Acțiunea României. 20 Sept. 1912—1 Aug. 1913, București, Imprimeria Statului, 1913

MURNU, GEORGE, Istoria românilor din Pind, Valahia Mare (980-1259), București, 1913.

MURNU, GEORGE, Pentru românii din Peninsula Balcanică, Cuvântare rostită în ședința Senatului, București, 1920.

NIEBUHR, B. G., Lectures of The History of Rome From the Earliest Times to the Fall of the Western Empire, London, 1850, Vol. I, The Pelasgians

PERTZ, GEORG HEINRICH, Monumenta Germaniae Historica. Anales Barensis, tom V

PEYFUSS, MAX DEMETER, Chestiunea aromânească. Evoluția ei de la origini până la pacea de la București (1913) și poziția Austro-Ungariei, traducere de N. Șerban Tanașoca, București, Ed. Enciclopedică, 1994.

POUQUEVILLE, F.-C.-H.-L., Voyage de la Grèce, Chez Firmin Didot, Paris, 1820.

REPORT OF THE INTERNATIONAL COMMISSION TO INQUIRE INTO THE CAUSES AND CONDUCT OF THE BALKAN WARS, Carnegie Endowment for International Peace, 1914

SCHLUMBERGER, G. L'Epopée Byzantine, Hachette, Paris, 1896

SCHMIDT, H. S. ET AL, THE HISTORY AND GENETICS OF THE AROMUN POPULATIONS. BIENNIAL BOOKS OF EAA 2000 1, 29–37.

SCHMIDT, H. S. ET AL: SOUTH BALKAN POPULATIONS, COLL. ANTROPOL. 27 (2003) 2: 501–506

SCHWANDNER-SIEVERS, STEPHANIE, FISCHER, BERND J.,
Albanian Identities: Myth and History, C. Hurst Se Co.
(Publishers) Ltd, 2002

STEFANOSKI – AL DABIJA, BRANISLAV, Scurtă istorie descriptivă
despre originea makedon-aromânilor (de la preistorie până la
colonizarea Daciei), Ed. CNI Coresi, București, 2011

TĂUTU, ALOISIE L., Devotamentul lui Ioniță Asan către Scaunul
Apostolic al Romei, 'Buna Vestire', nr. 1, 1966.

TRAD. POPA-LISSEANU, G., Izvoarele istoriei românilor, Vol. II,
Descrierea Europei Orientale de geograful Anonim, Tipografia
Bucovina, București, 1934

V. ARION, V. PÂRVAN, G. VÂLSAN, PERICLE PAPAHAGI ȘI G.
BOGDAN-DUICĂ, România și popoarele balcanic, Tipografia
românească, 1913

VULCAN, IOSIF, Emanuil Gozsdu, Revista 'Familia', Nr.6., Oradea,
1866

ZIARUL 'TIMPUL'

WEIGAND, GUSTAV, Die Aromunen. Ethnographisch-philologisch-
historische Untersuchungen, Leipzig, 1895.

WEB

Kahl, Thede, Aromanians in Greece: Minority or Vlach-speaking Greeks?, http://www.farsarotul.org/nl27_1.htm

A History of the Balkans since the Nineteenth Century, Imaging and Inventing the Balkans in Historiography, https://docs.google.com/viewer?a= v&pid=sites&srcid=ZGVmYXVsdGRvbWFpbnxyb2RhbnRoaXR6Y W5lbGxpYW5oaW5fGd4OjE5NzNjOTk0MDc0MDJjZTE

Encyclopedia Britannica, http://www.britannica.com/EBchecked/topic/1574478/Volcae

Wikipedia, http://ro.wikipedia.org/wiki/Wikipedia:Termeni_de_utilizare

Felipe Fernandez-Armesto, interviu în Pulse Berlin, http://www.pulse-berlin.com/index0985.html?id=146

http://ro.wikipedia.org/wiki/Torna,_torna,_fratre!

Anna Comnena, The Alexiad, BOOK X etc... http://www.fordham.edu/halsall/basis/AnnaComnena-Alexiad00.asp

Ehler et al, Y-chromosomal diversity of the ValachsY-chromosomal diversity of the Valachs from the Czech Republic: model for isolated population in Central Europe, CMJ, 2011, http://cmj.hr/default.aspx?ID=11764

A genetic atlas of human admixture history

Companion website for 'A genetic atlas of human admixture history', Hellenthal et al, Science (2014). http://admixturemap.paintmychromosomes.com/

Acu, Dumitru, Prof.univ.dr., Andrei Şaguna, Asociaţiunea ASTRA şi lupta pentru unitate naţională, http://www.tribuna.ro/stiri/ cultura/andrei-saguna-asociatiunea-astra-si-lupta-pentru-unitate-nationala-44738.html

Vlad, Valentin I., Acad., ASTRA şi cultura poporului român, p.1 www.acad.ro/com2011/doc/ASTRAacadVlad.doc

TRANSSILVANICA, Biblioteca digitală, Biblioteca Centrală Universitară 'Lucian Blaga' Cluj-Napoca, http://documente.bcucluj.ro/web/bibdigit/periodice/transilvania/192 5/ BCUCLUJ_FP_279996_1925_056_005_006.pdf

ILLUSTRATIONS

Figure 1 - Macedonia and Thessaly on the map of Greece, Historical Atlas by William R. Shepherd, New York, Henry Holt&Comp. 1911

Figure 2 - Great Vlachia, next to Epirus

Figure 3 - The Byzantine Empire, 1265. The Historical Atlas, William R. Shepherd, 1911.

Figure 4 - Macedonia and Thessaly on the map of Greece, Historical Atlas by William R. Shepherd, New York, Henry Holt&Comp. 1911

Figure 5 - Vlachs populated area in Moravia, Valašsko region, in the Northwest of the Carpathians, and in the nearby area

Figure 6 - Lumina, Popular Magazine of the Romanians of the Ottoman Empire - COVER

Figure 7 - Fragment from number 10 festival, October 1905 of Lumina, popular magazine of the Romanians of the Ottoman Empire „Publication of the educational and religious staff of Turkey", dedicated to the 25-th anniversary of the foundation of the Romanian High School of Bitolia, with a list of localities, with the names used at that time by their inhabitants.

Figure 8 - Pastor and boys from Lânga, Weigand, Die Aromunen Erster Band. p. 63, Digitized by Google

Figure 9 - Map of roman coloniae during the second century, Roman coloniae, (Wikipedia Commons)

Figure 10 - Romance-speaking Europe, (Wikipedia Commons)

Figure 11 - Albanian monks, Aromanian shepherd and merchant from Moscopole, Weigand, Die Aromunen Erster Band, p. 104, Digitized by Google

Figure 12 - Aromanian family from Pleasa (right an Albanian). Weigand, Die Aromunen Erster Band, p. 113, Digitized by Google

Figure 13 - Moscopole, Weigand, Die Aromunen Erster Band, p. 67, p. 113, Digitized by Google

Figure 14 - Aromanian family from Pleasa (right, an Albanian), Weigand, Die Aromunen Erster Band, p. 113, Digitized by Google

Figure 15 - Aromanian church in Moscopole nowadays, Photo: Eugene Matzota

Figure 16 - Neighbor-joining tree of the Balkan populations studied (four DNA-STRs and 19 classical markers)

Figure 17 - Geographic location of the samples analysed. Symbols represent the linguistic classification of the samples: Italic (stars), Slavic (circles), Greek (triangles), Albanian (square).

Figure 18 - Mount Athos, Photo: Eugene Matzota

Figure 18 - Epirus between 1205 and 1230, (Wikipedia Commons)

Figure 19 – Alexis I Comnenus and Hugues the Great (Wikipedia Commons)

Figure 20 – Epirus between 1205 and 1230, (Wikipedia Commons)

Figure 21 – Second Vlach-Bulgarian Empire, (Wikipedia Commons)

Figure 22 – The Monument of Ioniță Caloian in Varna

Figure 23 – De regno Dalmatiae et Croatiae, cover page, Digitized by Google

Figure 24 – De regno Dalmatiae et Croatiae, Chapter, Digitized by Google

Figure 25 - De regno Dalmatiae et Croatiae, Excerpt from the "About Vlachs" Chapter, Digitized by Google

Figure 26 – Dimitrie Cantemir

Figure 27 – Moscopole city crest The Makedon-Vlach Grammar, Digitized by Google

Figure 28 – The Makedon-Vlach Grammar, Cover page, Digitized by Google

Figure 29 – The Makedon-Vlach Grammar, Digitized by Google

Figure 30 - Orthodox church of Miskolc, The iconostasis

Figure 31 - Regions disputed by Balkan countries.

Figure 32 - Balkan countries' aspirations.

Figure 33 - Changes in the Balkans following the two conferences, one in London and one in Bucharest

Figure 34 - St. Nicodim of Tismana

Figure 35 - St. Joseph the New of Partoș

Figure 36 - Rigas Velestinlis

Figure 37 - Ioan Coletti

Figure 38 – Emanuil Gojdu

Figure 39 – Andrei Şaguna

Figure 40 – Ioan D. Caragiani and the founding members of the Romanian Academy

Figure 41 – Constantin Belimace

Figure 42 – Spyridon Lambros

Figure 43 – Aleksandër Stavre Drenova

Figure 44 - The Manaki brothers, Manaki brothers photo collection

Figure 45 - Poster from a Manaki brothers film, Manaki brothers photo collection

Figure 46 - Coat of arms – Mocioni

Figure 47 - Coat of arms - Mocioni de Foeni

Figure 48 - Gheorghe Simeon Sina

Figure 49 - The Sina Palais on a bigger map of Vienna, by Carl Graf Vasquez

Figure 50 - Nikolaus Dumba in his office in Dumbas Palace, Vienna